T0272217

Is computerised production transforming work roles, as recent debates about flexible specialisation and post-Fordist manufacturing suggest? This book focuses on the key case of metalwork batch production in Britain, Italy, Japan and the USA. Looking at technological, political and social developments from a comparative perspective, it suggests that comprehensive factory principles never fully replaced workshop organisation.

Drawing on empirical case studies of flexible manufacturing systems, Bryn Jones offers a new distinction between the bureaucratic bias of Taylorism and the product standardisation approach of Fordism, and questions whether computerised production is actually transcending Fordism. Instead of the often predicted models of deskilled, centrally controlled work, or a decentralised craft renaissance, he shows a greater likelihood of national variations and tensions between factory and workshop principles continuing into the contemporary age of computerisation.

Forcing the factory of the future

Forcing the factory of the future

Cybernation and societal institutions

BRYN JONES
University of Bath

PUBLISHED BY THE PRESS SYNDICATE OF THE UNIVERSITY OF CAMBRIDGE
The Pitt Building, Trumpington Street, Cambridge CB2 1RP, United Kingdom

CAMBRIDGE UNIVERSITY PRESS
The Edinburgh Building, Cambridge, CB2 2RU, United Kingdom
40 West 20th Street, New York, NY 10011-4211, USA
10 Stamford Road, Oakleigh, Melbourne 3166, Australia

© Cambridge University Press 1997

This book is in copyright. Subject to statutory exception and to the provisions of relevant collective licensing agreements, no reproduction of any part may take place without the written permission of Cambridge University Press.

First published 1997

Printed in the United Kingdom at the University Press, Cambridge

Typeset in 10/12½pt Plantin

A catalogue record for this book is available from the British Library

Library of Congress Cataloguing in Publication data

Jones, Bryn, 1946–
Forcing the factory of the future: cybernation and societal institutions / Bryn Jones.
p. cm.
Includes bibliographical references and index.
ISBN 0 521 57206 1 (hc)
1. Production management – Data processing. 2. Flexible manufacturing systems. I. Title
TS157J66 1997
670.42'7–dc20 96–44802 CIP

ISBN 0 521 5720 6 1 hardback

VN

To the memory of J. S. J. who would have understood.

Contents

Figures

Tables

Acknowledgements

The number of those who have directly or indirectly contributed to this book reflects the time it has been in preparation. For direct contributions from Britain, I must thank Peter Senker, my colleague Michael Rose and former colleague Harry Collins, for their advice and constant reminders. I owe a special debt of gratitude to Peter Scott for his unfailing help with difficult sources, and sharing details of his own research findings. In the USA Nathaniel Cook, Steven Early, Frank Emspak, Joel Fadem, Charles Ferguson, George Hutchinson, Gene Summers, Chuck Sabel and Jonathan Zeitlin gave invaluable advice, time and cooperation. Special thanks for the US coverage go to Ben Harrison and Maryellen Kelley, for not only sharing their knowledge and home, but also for their friendship. In Italy 'sono ingrado' to too many to thank individually. At the risk of under-estimating the assistance of others I would like to thank especially: Patrizio Bianchi, Sebastiano Brusco, Vittorio Capecchi, Claudia Ceccacia, Antonio Cocossa, Giovanni Contini, Angelo Dina, Maura Franchi, Francesco Garibaldo, Giuseppina Gualtieri, Sandro Maestralia, Angelo Pichieri, Vittorio Rieser ('per tanti assistenze'), Mauro Ronchetti, Michele LaRosa, Fabrizio Sarti, Mauro Selmi and Paolo Zurla. For information and hospitality without which my Italian visits would have been shorter and poorer my warmest thanks to Beppe and Lucia Della Rocca, and especially Roberto Benatti and Laura Zoboli. In Japan I must thank the scientific section of the British embassy in Tokyo, Yoshikuni Onishi, Jun Yamada and Sadayoshi Ohtsu. Last, but no means least, thanks to the ever-supportive Phil, and my children Leah and Asa who somehow managed to grow and thrive in the perpetual shadow of this study.

List of acronyms and abbreviations

AACP Anglo-American Council on Productivity
AEU Amalgamated Engineering Union (UK)
AUEW Amalgamated Union of Engineering Workers (UK)
BBC British Broadcasting Corporation
CAD Computer Aided Design
CAM Computer Aided Manufacturing
CIM Computer Integrated Manufacturing
CGIL *Confederazione Generale Italiana di Lavoro* (General Confederation of Italian Labour)
CISL *Confederazione Italiana Sindicalista Liberi (Confederation of Free Italian Unions)*
CNA *Confederazione Nazionale del' Artigianato* (National Confederation of Artisans)
CNC Computer Numerical Control (of machine tools)
CNP National Productivity Council (Italy)
DC *Democrazia Cristiana*
DNC Direct Numerical Control
DTI Department of Trade and Industry (UK)
FLM *Federazione Lavoratori Metalmeccanici*

	(Federation of Metalworking Workers)
FIM	*Federazione Italiana Metalmeccanica* (Italian Metalworking [union] Federation)
FIOM	*Federazione Impiegati Operai Metallurgici* (Federation of Metalworking Staff Workers)
FMC	Flexible Manufacturing Cell
FMS	Flexible Manufacturing System
FNAM	*Federazione Nazionale Artigiani Metalmeccanici* (National Federation of Metalworking Artisans)
GE	General Electric (US corporation)
GT	Group Technology
ICAM	Integrated Computer Aided Manufacturing
IUE	International Union of Electrical Workers (USA)
MAP	Manufacturing Automation Protocol
MIT	Massachussets Institute of Technology
MITI	Ministry of International Trade and Industry (Japan)
NC	Numerical Control (of machines)
PCI	*Partito Communisto Italiano*
TASS	Technical, Administrative and Supervisory Section (of British AUEW)
UAW	Union of Automobile Workers (USA)
UIL	*Unione Italiana del Lavoro* (Italian Union of Labour)

I The workshop versus the factory

1 Introduction: explaining factory evolution

The factory is the organisational and economic core of industrialism. It is a complex of planning, controlling and operating activities. Many believe: 'we have now the technology' to replace this complex of individual and social activities by an all-encompassing, integrated machine system. Micro-electronic computers can control and co-ordinate a variety of automatic equipment from a central source. Since the 1980s journalists, politicians, industrialists and technologists have linked new micro-electronic technology with the imminent arrival of 'workerless', robotised factories. Yet many of the principles underlying this 'cybernation'[1] have motivated factory development since the industrial revolution. Moreover, such integrated mechanisation of production is a goal that has hardly ever been realised. This book aims to assess the feasibility of these latest attempts; both in relation to previous failures, and the interaction between successive technologies and social and economic factors.

'Automation' refers to the age-old trend of replacing discrete human actions by machine operations. When first coined, in the 1940s, it referred to the linking of different machines by new controlling devices. However, in the last two decades technologists have envisaged automatic control of the broader execution, integration and control of entire functions and groups of functions – material processing, assembly, testing, storage etc. Because such a process implies the automation of automatons new terminology is useful. Thus this latter development, which concretises long-standing aspirations for a 'self-acting' factory, will be referred to in this book as 'cybernation'.[2] It will be argued that cybernetic, or self-governing control, is central and inherent to the logic of factory production. Since its consolidation in the nineteenth century the logic of the factory has meant more than aggregating different activities under one roof. Its perennial promise and attraction is the conversion of entire productive functions into co-ordinated machinery, or machine-like routines; characterised by regularity, predictability and precision. The fusion of cybernetics – the technology of control – with micro-electronic computing power is only the latest, most advanced attempt at perfecting this logic.

The credibility of this vision has enjoyed wide historical swings. Its recent ascendancy has been renewed by a variety of factors: intensified international competition, the crumbling of traditional Western factory techniques, Japanese productive disciplines, and the versatility of micro-electronic technology. Billions of dollars have already gone into cybernating factories. The wider economic and social stakes for employment, business profitability and the wealth of nations, are huge. The factory may encapsulate the organisational nature of industrialism (Perrow 1991, p. 744), but its spirit has become the goal of cybernation. Can this ethos be materialised at last; and, if so, on what social and economic pre-conditions?

Pursuit of the principles underlying cybernation is not new. So if it is a long-standing goal of factory organisation, then one must question whether today's superior technology is a sufficient condition for achieving the automatic factory. In the past, alternative organisational forms based on social qualities repeatedly, and successfully, rivalled earlier versions of cybernation. Gaps in the structure of mechanisation were filled in various ways: by squads of information-processing clerical workers, by craftworkers, by sub-contracting to smaller firms and, most notably, by falling back on the workshop model of production: a model from which some types of mechanised factory have, nevertheless, successfully evolved. The following account argues that, even with today's advanced technological resources, adoption of integrated mechanisation or reliance on socio-organisational arrangements depends on the specific and separate dynamics of social, economic and political events. Evidence for this proposition comes from both my own personal study of computerised production systems, and current social science debates about the broader trends in industrial organisation, automation and the role of human skills.

In these debates the actual adoption of new technologies of control and processing is not disputed. The key issue is whether this mechanisation continues, or departs from previous trends in the organisation of operations and use of workers' skills. Schematically, there are two main sets of views. One emphasises continuity in the aims and effects of the technology, and correspondence with related social developments. The other view sees the accompanying socio-organisational changes as relatively novel, semi-autonomous and discontinuous. Continuity approaches are dominated, but not completed, by traditional Marxist assumptions of an inherent logic in industrial change – promoting ever-greater simplification of tasks and centralisation of organisational control. From the discontinuity perspective, on the other hand, new production technologies are being accompanied by changed economic purposes and skill requirements: decentralised organisation and a more decisive role for production workers' tasks and skills.

Continuity perspectives linked to conventional Marxism see the developing division of tasks between workers and computerised machines as intensifying the century-old deskilling and centralisation of control promoted by F. W. Taylor and scientific management (Braverman 1974). Viewed as a discontinuous development, on the other hand, the adoption of computerised production technologies is seen now as subject to new industrial priorities: more variety in product specifications, more design changes, and decentralisation of production away from large integrated factory operations. These priorities, it is argued, render irrelevant the technology's potential for centralisation of control and substitution of skills. They require, instead, more innovative reprogramming of the systems, extra and changing product and process specifications, and thus skilled workers with more autonomy. The versatility and flexibility involved are allegedly both incompatible with Taylorism, and also part of a broader, concurrent shift away from the mid-twentieth-century system of 'Fordist', or mass production. Hence the key issues in this debate inform our concern with the specific logic of cybernation: what is, or was, the scope and significance of 'Fordism', is it already, or soon to be transformed into 'Post-Fordism', and what is its relationship to Taylorist centralisation of control?

The Italian Marxist leader, Antonio Gramsci conceived 'Fordism' to describe the essence of US industrial practice. The concept entered recent debates about the reorganisation of the factory through its use by French 'Regulation' theorists, such as Aglietta and Palloix (Aglietta 1979; Palloix 1976), to define the mid-twentieth-century expansion of capitalism as a 'regulated' co-ordination of mass production with mass consumption. Businesses, such as the eponymous Ford Motor Company, allegedly prospered by large-scale standardisation of products and production methods. These also cheapened consumption goods and expanded workers' purchasing power, employment and wage levels. In conjunction with governments' income maintenance policies, mass production/consumption raised the overall level of demand. It should be noted that the all-inclusive sweep of this theory also places its accompanying image of the 'Fordist' factory – bigger, more centralised, with hierarchically simplified job tasks – within the institutionalisation of Taylorist work organisation. In these respects the Regulationists' thesis is consistent with the assumptions of a continuous view of mechanisation. It is on the question of what is now succeeding Fordism that crucial inconsistencies appear in this apparently clear-cut picture.

Regulationists claimed initially that Fordism was only evolving into a modified neo-Fordism. On the production side this new system was said to use computer technology to improve higher-level control over the design, planning and final production of goods (Aglietta 1979). Enhanced central control paradoxically allows some relaxation of the tight and narrow job tasks at the point of production. Repackaging of lower-level tasks into broader and more fluid work roles, with an ethos of job enrichment and collective responsibilities, allegedly loosens Taylorist specialisation in rigid job profiles. Computerised centralisation improves control, precision and flexibility, and hence cost effectiveness, and so boosts stagnating sales. In this interpretation the simplified-liberalised task repertoires also reduce workforce grievances and help to minimise any remaining problems at the point of production (Aglietta 1979, pp. 126–30). Thus Regulationists' broader view of industrial change does not necessarily lead to a radically different conclusion about contemporary production work than that of more traditional Marxism. Others, however, perceive a more fundamental metamorphosis of Fordism, seeing a crucial discontinuity between traditional Fordism and the use of new technologies and workers' skills.

The key argument of this latter view is that the standardised mass-production and steady and homogeneous market demand of Fordist industry are inter-dependent. When these do not coincide alternative, less rigid types of operation may flourish. Sabel and Zeitlin point out the historical evidence of superior performance by smaller-scale, craft-oriented operations matched to smaller and more varied markets. Brusco and Sabel highlight the successful expansion of this 'flexible specialisation' through networks of small artisanal firms since the 1970s to fill the vacuum left by the weakening capabilities of large Fordist firms in Italy: an expansion in which traditional craft skills are enhanced by more piecemeal use of computerised equipment. Piore and Sabel generalised from this small-firm pattern to suggest that flexible specialisation is a more widespread and rival model to Fordism, even amongst some large firms. It concentrates on short runs of rapidly redesigned products, catering to shifting consumer tastes, and segmented and unstable demand. The crucial element in this new production and market nexus is, they claim, a new 'technological paradigm'. This is an entrepreneurial and managerial conception of technology which breaks decisively with Fordism and also offers a more realistic opportunity for industrial renewal than the compromise neo-Fordist adaptation identified by Regulationists (Sabel and Zeitlin 1985; Brusco 1982; Sabel 1982; Piore and Sabel 1984; Murray 1988).

Detailed assessments of specifically technological changes in manufacturing have proliferated in the last twenty years. Many of these are, of course, largely unconcerned with issues of historical continuity or discontinuity. Some emphasise both continuity and the potential for discontinuity. Zuboff (1988) argues that the latest phase of computerising production is, in some respects, a continuation of earlier phases in the transfer of skills from human workers to organisational and machine systems. On the other hand, she claims that integrated computerisation creates distinctively new 'informating' capabilities and skill requirements rather than simply replacing existing human–machine interactions. Significantly, however, Zuboff admits that authority systems, managerial beliefs and attitudes and the organisational prerequisites of firms and factories, combine to frustrate the full realisation of the employees and firms using these 'informating' capacities (Zuboff 1988, pp. 251, 299–305). As Zuboff partly concedes (pp. 403–4) we have to look beyond the technical organisation of production to wider social institutions, such as industrial relations, to understand the conditions facilitating or restricting cybernation.

The debate between the discontinuous and continuous perspectives provides just such an entrée into the broader explanatory contexts of the latest evolution of factory organisation. If the current transition is a continuation of key tendencies within Fordism then its business strategies, and organisation of operations, ought to accelerate a move to a cybernated system of production; perhaps with some cosmetic autonomy for the few workers doing the remaining peripheral tasks. Contrarily, however, if the discontinuous view is correct, and flexible specialisation has displaced Fordism, then industrial efficiency is not likely to be sought through cybernation. Thus establishing the relative prospects of small-scale, internally or externally decentralised production on the one hand, or cybernation on the other, will also clarify this continuity–discontinuity debate on the future of the factory.

The pathology of batch production

The empirical focus of this book is past and current factory organisation in metalworking industries. Metalworking, at the heart of all manufacturing industry, is crucial to industrial economies. It is also a critical test case for both the logic of cybernation, and the debates over the nature and future of Fordism. Cybernation has largely taken over in bulk processing industries, such as petro-chemicals, where different operations on indivisible flows of materials can be simultaneously executed (Hirschhorn 1984, pp. 43–5). The last outposts of batch-craft operations in raw material processing, such as pulp-paper plants, have recently been converted to integrated computer-controlled systems (Zuboff 1988). Most metalworking production, however, still exemplifies the complexities of batch production. The numbers of a particular product or component required are small, and the operations required to fabricate them are numerous and sequential. Changes to the designs and settings of machinery require repeated and sometimes unpredictable applications of skill. The history of metalworking is one of frequent and varied innovations in mechanising aspects of small-batch production. Yet, until recently, many operations were composed of discrete concentrations of considerable manual and intellectual skills. Micro-electronic cybernation of this most challenging of industrial battlefields, would seem to assure its final victory.

Batch metalworking has a related significance for the debate over the continuity or discontinuity of Fordism. Fordism, in the sectors of metalworking which escaped the constraints of batch production, meant pre-micro-electronic forms of automation and standardising products and processes. Chapter 1 analyses the conditions for, and restrictions on, this breakthrough into mass production at the early twentieth-century Ford car factories. Metalworking batch production was also the initial proving ground for F. W. Taylor and the Scientific Management movement. Taylor himself drew on his early work experience as a metalworking machinist. Yet both Fordist and Taylorist methods have also been stymied in most metalworking operations.

The concentration of the early Regulationists, and then the advocates of Flexible Specialisation, on an alleged crisis of mass manufacturing, is misleading. The predominance of batch production in metalworking, and the limited application of Fordism there, suggests a different kind of crisis and other kinds of questions.

- Is Fordism synonymous with mass production?
- How, and how far, did automobile and other putative mass producers 'escape' the constraints of batch production?
- Why was much of the rest of metalworking relatively unsuccessful in applying Fordist methodology?
- Can micro-electronic automation provide the missing ingredients?
- Can cybernation or integrated automation be pursued within batch production if Fordism or neo-Fordism are inapplicable there?

Fordism and Taylorism: analytical and historical distinctions

The following chapters will develop answers to these questions. Chapter 2 traces the logic of cybernating production back to the industrialisation of metalworking at the beginning of the nineteenth century. It confirms the claims of the Flexible Specialisation thesis: that the routinisation of production was limited by the external variability of demand and the internal variety of tasks and skills. However, examining the innovations in standardised production which preceded, and then transformed, automobile production shows that the Fordist production paradigm is more than mass production by specialised machines. Its distinctive features are: advance specification of component production of product parts and production processes by standardised designs and plans; as well as integration of the maximum number of processes on a 'flow of work' principle.

Chapter 2 argues that this lateral, or product-driven, standardisation is distinct from scientific management's discrete logic of *vertical, administrative controls* focused on specific tasks. Contemporary developments in both computerisation and so-called 'lean production' systems illustrate the distinctiveness of these dimensions. In principle, there is simultaneous computerisation of machining operations, transfer devices and inventory – what I call *horizontal integration* – and of hierarchical, vertical, controls over costing, staffing and production planning control functions. In practice the two domains are being computerised separately. The technologies for linking throughout each dimension are new; but the underlying aims are almost as old as factory production itself. Similarly the currently fashionable 'lean production' system (Womack et al. 1990) in the automobile industry also indicates the distinction between the Taylorist focused dimension of hierarchical control and the lateral, process dimension. Lean production organises consecutive operations so that one 'pulls' the necessary supply from its predecessor 'just-in-time'. It is aimed at shifting initiative from the vertical to the lateral axis.

Distinguishing Taylorist from Fordist measures thus clarifies their scope and impact, and more recent attempts to rationalise batch production.

These distinctions are necessary because, like well-worn coins, habitual use of the terms Taylorism and Fordism has tended to efface their distinguishing features. For example, in the debate over flexible production successors to Fordism, protagonists largely follow Piore and Sabel's picture of Taylorism – as a work organisation philosophy analogous to, and consistent with, Fordism's engineering of standardised mass production processes (Piore and Sabel 1984, pp. 126, 142; Hyman 1988; Wood 1988, 1989). By contrast the labour process school, spearheaded by Braverman, identifies Taylorism with the definitive principle of industry in modern, monopoly, capitalism, as: 'the explicit verbalisation of the capitalist mode of production'. Others, however, credit Fordism – the manufacturing system developed by Henry Ford's automobile firm – with subsuming Taylorism's particular techniques within a more comprehensive system of operations management and business strategy. It is argued, for example, that Ford himself:

> took over the essential aspects of Taylorism ... separation of design and innovation from execution, division and sub-division of jobs, each movement allowed a specific time (Palloix 1976, p. 59).

Or, that rigorous application of Taylorist task fragmentation reached new heights in Fordist mass production (Zuboff 1988, p. 43). On this view its fuller realisation needed: 'mass markets, mass production and velocity of throughput' (Littler 1983, p. 14).

An alternative interpretation is that only certain aspects of Taylorism, such as time and motion study, were appropriate for Fordist cost controls, on the grounds that these aspects are only subsidiary techniques that can be supplemented, or replaced, by other non-Taylorist methods. In this view Fordism is seen as the combination of both minimised task ranges for individual workers and highly specialised product and process technologies for the manufacture of high volumes of standardised products (Blackburn, Coombs and Green 1985, pp. 43–5). Others (Williams et al. 1987; Williams, Haslam and Williams 1992) have rejected the entire conception of Fordism on the grounds that it lacks corresponding empirical examples. I wish to argue that Taylorism and Fordism are empirically verifiable, but that their conceptual distinctiveness must first be redefined.

Taylor's emphasis was, as Kelly points out, on the 'scientific' analysis and repackaging of tasks, rather than their necessary fragmentation. Thus the most economical pattern of work roles could mean – according to Taylor himself – *augmenting* the task range of some jobs, and not de-skilling, through task fragmentation. Always provided that the reorganisation was based on prior, scientific, analysis and design by management (Kelly 1982, pp. 23–4). Predetermination of the fine details of a uniform product, however, made deskilling integral to Fordism. Seen from this perspective the overlap between Fordism and Taylorism is reduced.

A common aim of both systems was to increase the output of the individual production worker. But there the similarity mainly ends. All of the Taylorists' analysis and reorganisation was geared to individual productivity as the focal point of enterprise efficiency. For Fordists, however, the productivity of the individual worker was an intermediate goal, a means for making a uniform product by uniform standards cheap enough to sell to a mass market of unknown consumers. To overstate slightly: the overriding goal of Taylorism – routinised, non-conceptual tasks – was largely a by-product of the more comprehensive business strategy of Fordism. Those who stress the similarity of Fordism and Taylorism are correct inasmuch as the two approaches overlap empirically. Yet they apply largely to distinct dimensions of production organisation. Taylorism was geared to the vertical, or control, dimension – through which organisations hierarchically regulate product specifications, and quantities of input and output. The main focus of Fordism is the horizontal, or lateral, dimension of production organisation, that concerned with the process by which the energy, raw materials, and components are made into a finished product.

Chapter 2 also contrasts the receptiveness of US industry to Taylorism and Fordism with their initial failure in Britain. This contrast continues in Chapter 3's overview of their import into Britain and Italy after the Second World War. Differences in their adoption between Europe and the USA confirm the analytic separation of Taylorism and Fordism. Contrary to some claims it appears that the take-up of Taylorist practice – as opposed to Taylorist doctrines – was more systematic in the UK after the Second World War, than during the inter-war years. Moreover Tayloristic techniques, mainly work-studied, incentive payment systems, were adopted in preference to pure Fordist methods – which were deemed either technically or commercially inappropriate. In some respects Fordism, as a technical strategy, only gathered momentum in Britain and Italy as the localised versions of Taylorism began to fail. In the UK in particular, the rigour of Taylorism was weakened by the influence of national industrial relations institutions.

Subsequent chapters deal with the continuist view of Fordism that a coincidence of micro-electronics applications necessarily appeared at the same time as more detailed interest in Fordist techniques. However, the related Marxist thesis, of a labour process in which industrial change means a progressive centralisation of control and deskilling of work, is not separately analysed here. Its claim has been critically assessed elsewhere.[3] Chapter 4 does examine its possible relevance to the continuity thesis in metalworking production, in the undoubted involvement of Taylorist aims in the diffusion of individual, computerised (NC) machine tools. However, chapters 3 and 4 show that the pursuit and realisation of Taylorist aims are far from being universal or successful. Specific social and political conditions influence how businesses and policy makers perceive their relevance and attempt to realise them. Moreover, the appropriate production opportunities will almost certainly vary between times, modes of dissemination, and favourable or adverse socio-political and economic conditions as between different countries.

Braverman-style deskilling theories reduce both the practical and technical complexities and the social and political conditions to constituents of an alleged all-encompassing economic logic. Yet, work organisation varies between societies because of complexes of socio-economic and political institutions that are specific to those societies (Maurice et al. 1986). Thus a uniform and self-fulfilling logic to industrial organisation is implausible. Such a critique shows the unsustainable reductionism in this kind of approach, but it is not only this Marxist labour process theory which reduces industrial-organisational change to a single, implausible cause.

Social and technological reductionism: trajectories and paradigms

Another form of reductionism is technological determinism: the belief that key socio-organisational features necessarily follow from the successful introduction of a new technology. Although *persona non grata* in social science discourse it is still assumed elsewhere that technology determines social and organisational aspects of production (e.g. Scott 1986, pp. 225–7). Simply condemning technological reduction, however, does not explain its role in industrial and organisational change; and if technology had no effects then managements would hardly bother to introduce it (Clark and McLoughlin 1988, p. 100; Winner 1977, p. 75).[4]

If technology is completely discounted there is a danger – exemplified in the problems with the labour process thesis – of technological determinism being replaced by economic or social determinism. The notion of technological trajectories is one way to avoid such reductionism. In this approach long-run industrial change is periodised partly as the rise and fall of specific types, or complexes, of technology: steam power, the internal combustion engine, micro-electronics and so on (Dosi 1982). Piore and Sabel similarly talk of 'zig-zagging paths' and 'branching trees' of technological development (Piore and Sabel 1984, p. 39). Such metaphors are a way of allowing some autonomy to the internal logic of technology. But what pushes the technology on to a particular trajectory or shifts it on to new paths?

One explanation of why a technological pattern becomes locked into one line of development is the concept of a technological paradigm (Piore and Sabel 1984, p. 44; Perez 1985). Like Thomas Kuhn's original idea of scientific paradigms, this is a mutually reinforcing set of beliefs held by practitioners. The paradigm inhibits the acceptability of alternative perspectives and guides further development within the same framework. The following account proposes the slightly different relevance of production paradigms,[5] because the adjective 'technological' may be misleading. The intellectual frameworks of industrial managers and production engineers do not extend to a familiarity with all the current principles of the leading technologies. On the other hand, the cognitive frameworks of these actors do include non-technological elements of production: cost accounting criteria, skill requirements, organisational procedures, and so on.

However, if paradigms are ultimately rooted in beliefs about what is appropriate and feasible, then what disturbs these roots, and exposes the possible relevance of alternative paradigmatic principles? Kuhn's scientists work in relatively closed social worlds. Industrial practitioners, on the other hand, are exposed to a wide variety of socio-economic and political influences: new theories of business organisation, the industrial policies of the state, definitions of new customer preferences, the changing institutions and practices of employment and labour relations. Thus a range of socio-economic and political influences will influence the assumptions and premises in the production paradigms held by engineers and managers. In some circumstances the latter are particularly forceful or critical. For example, in the USA the combined effects of egalitarian consumer demand, and masses of cheap and under-unionised migrant labour, weakened the paradigm of craft-engineered luxury goods production and stimulated that of Fordism. But doesn't making production paradigms a response to social forces raise again the risk of social or economic reductionism? One way of avoiding reductionism is to regard the socio-economic and political influences as trajectories, in like manner to the technological paths.

Here a fuller definition of the nature of trajectories is needed. If it is helpful to think of technological, organisational, labour relations and other paradigms and complexes as trajectories, then they must be semi-autonomous. They must have their own internal logics and institutions. While mutually co-existing for long periods, they may also intersect. At these critical junctures, their different characteristics may be either conflictual or complementary. Chapter 3 provides a major example of the latter possibility. In Western Europe, after the Second World War, the political trajectory of Cold War international relations and state intervention championed the production paradigms of Fordism and Taylorism through the technical assistance programmes of the US Marshall Plan.

Readers with long sociological memories may recognise in this conception of semi-autonomous trajectories the sadly neglected, now unfashionable, ideas of the late Louis Althusser. His disaggregation of any specific historical 'stage' into its constituent economic, political and ideological processes remains analytically invaluable. It showed that such processes have their own distinctive 'times' and rates of development whose periodic interactions are complex and destabilising – precisely because each process has developed at its own pace and with its own logic (Althusser and Balibar 1970, pp. 99–105, 292–302). A full explanation of this approach is beyond the scope of the present study; but its implications are apparent in much of the substantive exposition. The conjunction of trajectories, whose histories and significance we will be trying to disentangle, concerns computerised production operations: what contemporary technologists refer to as 'computer-integrated manufacturing', social scientists as 'flexible automation', and the business press as the 'factory of the future'.

National comparisons: sharpening the focus

National comparisons also help to distinguish between such paradigms as Fordism and Taylorism, and to avoid collapsing the different tempos of technical, socio-economic and political change into one homogeneous development. Piore and Sabel's 'Second Industrial Divide' between Flexible Specialisation and Fordism is based on national comparisons of developments in the USA, Germany, France, Italy and Japan. Such broad comparisons are more revealing than the assumption of a supra-national logic of capitalism; but they may still over-generalise. The breadth of their comparisons are made more problematic because of the simplicity of their industrial categories. Piore and Sabel equate 'mass production' with most American industry and, to a lesser extent, with post-war France. They then contrast it with the German, Italian and Japanese economies. In these latter countries, mass production did not, allegedly, successfully replace the alternative 'craft production' model. Yet Piore and Sabel ignore, for example, the case of twentieth-century Britain. Perhaps because it does not fall unambiguously into either the mass or craft production categories? Either it is an exception to their rule or the categories need re-examining.

The following chapters offer a more detailed focus on a narrower field. By concentrating on metalworking the range of variables is more manageable. The comparisons of Britain, Italy, Japan and the USA are selective and heuristic rather than comprehensive. Britain is the principal focus because its longer history of industrial change permits a more extensive archaeology of the different strata of production organisation. The USA is self-selecting as the heartland of scientific management and Fordism. Italy and Japan represent apparently contrasting developments away from conventional Taylorism and Fordism using different applications of computerised production technology. Italian applications are based on a radical divergence between large-firm neo-Fordism and small-firm flexible specialisation. Japan stands out for the blend of worker responsibility and advanced technology in neo-Fordist larger firms, but an exhaustive study should have included other distinctive national cases: Germany or Sweden being prime candidates. But the extra volume of data involved in adding these countries could only have been included by expanding the conceptual framework or restricting the depth of evidence.

The exposition introduces the national cases as their industrial developments become significant for the further development or frustration of Taylorism, Fordism and the logic of integrated automation. Thus the early chapters focus on Britain and the USA, with Italy then introduced to assess the impact of Taylorism and Fordism, and again in chapters 5 and 9 in the comparison of computerised automation in large Fordist-oriented plants with artisanal flexible specialisation. In chapter 9 Japan is discussed as a self-contained case of evolution towards cybernation (chapter 8). Similar case study evidence on the same evolution – through allegedly neo-Fordist Flexible Manufacturing Systems – is put into the national contexts of Britain and the USA in chapters 5, 6 and 7 respectively. Chapter 5 also begins the examination of the workplace interface of technological and socio-political trajectories through the nature and extent of workers' and managers' 'job controls' over task organisation in the operation of technology.

The genealogy of production organisation: workshop and factory

The closer focus on particular sectors and processes of production, such as batch production in metalworking, also provides finer distinctions than that between craft and mass production. The crudeness of craft vs. mass production – one term relating to skill level, the other to product volumes – reveals the need for an analytical history of the factory, as a socio-economic institution.[6] That exercise would take us far beyond the logic of cybernation; but, to understand that process more precise organisational distinctions are both possible and necessary.

Consider the two polarities of factory operations. The acme of contemporary industrial plant is the integrated, centrally controlled operations found in 'process' industries in petro-chemical operations. Other manufacturing sectors aspire to the principles of these 'flow' or continuous process complexes. As later chapters show, flow line levels of automaticity are one of the aims of computer-integrated mechanisation in batch production. There is also a direct line of descent between these integrated processes and the first factories of the Industrial Revolution. Although these often sprang from the opposite polarity of the workshop, they appeared first on a widespread scale in British textile mills. The central powering of looms and spinning machines by steam engines enabled Marx to exaggerate the textile factory as: 'An organised system of machines' because '. . . motion is communicated by the transmitting mechanism from a central automaton'.[7]

Marx, other contemporary observers, and even some recent historical accounts confused control through the transmission of power with control through independent information-processing mechanisms: what would be recognised today as cybernetic systems. Nevertheless, some of the early factory systems were based on rudimentary notions of integrated automation; a kind of proto-cybernation. Chapter 2 discusses these attempts. It confirms historical studies (Lazonick 1979), which show that in practice even the constituent machines still required inputs of human skill and judgement. Nevertheless the semi-integrated nature of the machinery in early nineteenth-century textile factories was closer to the flow principles of later mass production and process industries, than the workshop forms which previously predominated.

Full or quasi-integration of the machining processes in factories serves to simplify and dictate the remaining human tasks. The human tasks, however skilled, fill in the interstices between the capabilities of the machines. Moreover, the dominance of machinery processes, means that the closer human actions are to the machine systems, the more they tend to resemble repetitive, machine-like, supplements to their functioning. The human inputs and their social organisation are dedicated to furthering the integration of the machine systems. In the workshop, on the other hand, the dominant process is the social organisation of the human tasks.

The discrete, individual machines supplement this network of human functions. Whereas the factory is preoccupied with continuous operation of all its expensive machinery, in the workshop the machines function as so many different tools – to be applied to specific jobs, as and when necessary. In metalworking, as in some other sectors, the machines for the true factories of the Industrial Revolution period were themselves made in workshops for most of the nineteenth century (Landes 1969, pp. 118–21, 183).

Flow principles are more easily maximised with large batch or mass production. Yet it is not mass production itself, but its combination with Fordist specialisation of products and processes which simplifies and routinises most tasks. Fordist organisation allows the more complex design, planning and machine setting activities to be hived off to separate functions and locations; while the actual production and co-ordination follows repetitive cycles of identical routines for each operation and product unit. By contrast, in the workshop scale of production is limited and the range of current products is often mixed and variable. The range of individual workers' tasks is more extensive: both horizontally – between different tasks at the same levels of skill – and vertically – between tasks at higher or lower skill levels. The correspondence between variable products and varied tasks makes jobs, that is work roles, more difficult to simplify and standardise. Yet, as chapters 2 and 3 show, Taylorist task control and simplification were an influential historical attempt to standardise workshop practices without necessarily standardising products and processes.

Many metalworking factories have remained, in key respects, collections of workshops under one roof. While the true factory is, as a whole, more suited to mass production, to large batches, or large units requiring many identical, or similar components, the workshop, on the other hand, is more appropriate to smaller production batches. However, the size of firms based on factory operations has made the latter appear as the commercial paragon. This factor, and the dominance of its production paradigms in engineering and management discourse, have often encouraged workshop businesses to emulate the prevailing principles of integrated factory production. Batch metalworking production stands on the cusp of these two principles of organisation. Where the scale of the product or the numbers of product units is large enough, then the factory model, and the logic of cybernation, may be attempted. In other circumstances workshop principles persist.

As the flexible specialisation hypothesis proposes, this distinction is not limited to survivals of a pre-factory era. Every day, all over the industrial world, the economics of batch production require manufacturing managers to decide whether to 'make or buy' metal components. Should they make a part with existing or new equipment within their factory? Or should they purchase its production from a sub-contractor; often a 'jobbing shop' making, mainly, relatively small batches for larger customers? In metalworking, unlike other industries such as texiles or footwear, workshops have not retreated to a luxury product fringe. Instead, there is a constantly varying frontier between the two regimes. At one time, perhaps when a production paradigm such as Fordism is ascendant, the forces of factory production will take over product lines and establishments from workshop units. In other instances, socio-economic and political developments – perhaps aided by market, product or technical changes – will reassert the commercial relevance of craft-based workshops, on territory otherwise seemingly more favourable to the factory.

Mass production is a species of the factory model and the workshop is the genus which contains craft production. Yet, as the metaphor of the variable frontier indicates, mass production in metalworking contains residual elements of 'craft production', and the craft methods of workshop organisation are potentially transformable into mass production's factory principles. The early factories, Fordism, and now the neo-Fordism of computer-integrated automation, represent successive attempts to break with the legacy of the workshop in batch-production metalworking. The following chapters analyse the dynamics of this apparently endless struggle. They also show whether there is continuity in a trajectory to the cybernated factory, or whether the 'end of Fordism' or an 'end of divided labour'[8] entail a restoration of the workshop.

Forcing the future

One of the most influential ideas in discussions of changing industrial organisation is the 'factory of the future' theme. Current technological reorganisation is judged in terms of its proximity to a fully developed model of computer-integrated automation of all the constituent activities of a manufacturing plant. I have called this the cybernation of factory production because it seeks to complete automaticity and eradicate the remaining elements of workshop practices. When he invented the term 'cybernetics' to describe the new science and technology of systems control, Norbert Wiener admitted that its principles of information feedback and automatic controls, extended back to the 'steam engine governor' of initial mechanisation (Wiener 1948, p. 43).

The vogue for flexible automation arises from the unification of servo-mechanisms, giving Wiener's adaptive control to individual machines, with the information-processing powers of micro-electronic computing. The latter make possible adaptive control across groups of machines and, in principle, throughout the entire factory.[9] This is a qualitative leap from the automation of disparate functions of power, material transfer and material transformation. Rather than add another adjective to the term automation, it seems more appropriate to acknowledge that this development is the automation of automation: the cybernation of the total of the factory's constituent 'automatic' operations.

From the viewpoint of factory logic the virtue of this development is that changes in product characteristics and production processes can be achieved with virtually complete automaticity. People undertake some high-level reconfiguring of the software; but routine physical and mental inputs of lower and middle-level workers are reduced almost to zero. This image has acquired an enormous cachet in some quarters. Its origins, internal logic – on the borders between science fiction and science fact – and North American constituency have been vividly portrayed by David Noble (Noble 1984, ch. 4). The prospect of automatic programming changes is also believed to abolish the mass-production/batch production distinctions which previously limited cybernation of the production operations to particular phases of mass production. The constituent machining, transfer, planning, and supply systems are automatic, and they are automatically integrated by computer systems which can reprogram the entire complex for different products and operations. This Computer Integrated Manufacturing (CIM) approach, whose most practical application is in production units known as Flexible Manufacturing Systems (FMS), offers Fordist-style automaticity without the requirement of mass-production scales of standardised products.

The perfection of cybernetic factory operation as a demonstrable technology, gave rise to the excited claims of business and engineering pundits. Yet the full-blown, contemporary realisation of CIM is almost non-existent. Even publicists of 'future factories' concede that different stages of preparing and executing production have nowhere been completely integrated by computer systems. Likewise it is acknowledged that workers still make some, albeit reduced, contributions to the continuity of production. Nevertheless, it is claimed, such imperfections in the future model are temporary expediencies. The blueprint of the completely reprogrammable, computer-integrated, automatic factory is, allegedly, perfectly precise. All that is needed for its realisation is the passage of time, and the ironing out of a few technical details.

These, and related predictions of epochal shifts in the broader economic and social structure of the late twentieth century all suffer from a logical deficiency that I shall call 'futurism', a style of analysis which is a close cousin of 'utopian' social theorising. In the latter tradition – exemplified by certain varieties of Marxism and neo-liberal advocates of free market economies – the significant social, economic and political features of the present are identified as those which constitute the seeds of a potentially new, and morally superior, social regime of the future. In this way, trade unions, co-operatives, joint stock companies, and the large-scale factory itself, have been seen as the forerunners of more advanced social forms of post-capitalist society. Futurists, such as Alvin Toffler and some advocates of a Post-Industrial social structure, share a similar logical approach; albeit without necessarily giving moral approval to the coming transformation. In their accounts the future is already microcosmically formed in the present. The general outline of what is to become the norm is already clear from the future work lives, life styles, industrial and social structures burgeoning today in allegedly more advanced locations, such as California and computer companies.[10]

A note on methods

Deciding which types, and how much evidence, to deploy in explaining and clarifying a complex social or economic development presents most social scientists with a number of dilemmas. Data from the author's single or connected empirical investigations – as a research monograph – show detailed causes and effects; and provide, hopefully, unambiguous answers to precise questions. However, when the question is broader and when the data are not so readily discoverable, then the social scientist must have recourse to a more discursive approach: interrogating and blending the evidence from a variety of other authors in order to construct a fuller picture. We are concerned here with the origins, evolution, and comparative variation of technological, industrial, social and political trajectories, all of which play a part in the formation of contemporary programmes of factory cybernation.

This research approach involves an interlinked compendium of both targeted empirical studies of selected cases of automation and cybernation, interviewing of expert figures, historical enquiry, documentary analysis and reportage from secondary texts. The contemporary empirical data were collected from successive investigations in Britain, USA, Italy and Japan between 1979 and 1990; mainly from case studies, by myself and post-graduate researchers, which involved some 500 interviews in total.[11] The sources for the historical material in chapters 2 and 3 were contemporary documents, secondary accounts and a limited amount of personal archival research. In addition to interview material from the personal visits to British, Italian, Japanese and US plants, technical reports, conference papers, and contractual documents were consulted in each society; as well as published academic studies from those countries.

So the general methodology is analytical, comparative and historical. The primary material from the factory interviews is embedded in a much broader account, which utilises the advantages of a comparative sociological approach, and an attempt to refine and utilise new theoretical distinctions. The clearest models for this approach are the 'societal effects' perspective of Maurice and his colleagues at the Laboratoire d'Economie et Sociologie du Travail, and the panoramic, socio-historical account of Piore and Sabel in their *Second Industrial Divide.* While the works of the former school (cf. Maurice et al. 1986) clearly surpass the present study for rigour and descriptive precision, it is hoped that it does not betray the insistence of the Aix group on the institutional specificity of industrial phenomena within paricular countries. Likewise Piore and Sabel's attempt to weave economic and socio-political trajectories together in a multinational overview stands as a partial model for my study. The reader can judge the success of this synthetic approach. But it must be pointed out that its four-nation comparisons and concentration on a single, crucial, industrial sector is both more ambitious and more focused than these paradigmatic predecessors, or more technologically focused monographs such as Wilkinson (1982), Noble (1984), Shaiken (1984), Zuboff (1988), Giordano (1992) and Whittaker (1990).

Outline of the thesis

The preceding analysis leads to three main questions. Will computer technologies realise factory principles of cybernation in the 'workshop' sphere of batch production? Is the emerging organisation a continuation of earlier industrial revolutions, or is it qualitatively different? Are the newer practices sufficiently powerful and universal to transcend the national variations that have developed in the past? The international evidence in the following account will suggest neither the final triumph of Fordism, via cybernation, nor an all-embracing resurgence of workshop-based flexible specialisation. For the foreseeable future a divergent production pattern is most likely: between a revitalised and computer-assisted workshop sector, and more highly mechanised Fordist practices. The fullest realisation of the former is likely where socio-political conditions allow networks of social institutions to support a class of neo-artisanal workers and owners, as in certain Italian regions.

However, the nearest thing to a cybernated form of Fordism in batch production, in one sense a 'neo-Fordism', is most likely where the institutions favour labour relations and work organisation similar to larger Japanese firms since the 1950s. The latter's compromise with the realities of socio-political conditions probably constitutes the furthest that large-scale industry can modify Fordism. Countries such as Britain or the USA, display a third variant: Taylorist, quasi-Taylorist, or 'workerless' production paradigms sustain policies of computer-integrated cybernation. Because Taylorism contradicts some aspects of Fordism, these characteristics are partly responsible for the present limitations of computer-integrated automation. However, because each of these Euro-American approaches *is* a paradigm, in the classical sense, some further *adherence to the goals* of cybernation in the north-western industrial world can still be predicted.

2 Past production paradigms: the workshop, Taylorism and Fordism

> Modern Industry had therefore itself to take in
> hand the machine, its characteristic instrument
> of production, and to construct machines by machines.
>
> Karl Marx, *Capital*, vol. I, p. 384

Introduction

Beneath forecasts of the revolutionary cybernation of factories are assumptions of historical precedents. Notably, that manufacturing evolution proceeds through systematic transformations: the mechanisation of the Industrial Revolution, Taylorism/Scientific Management, Fordism. An initial problem in assessing the validity of such characterisations is deciding what qualities define the dominance of one system of production rather than another. Unfortunately, the obvious criteria such as the preponderance of different production techniques, occupations, machines or organisational structures are unreliable guides. A contrast in early nineteenth-century Britain between the numbers of self-acting mules and the numbers of spinning-jennies points, unequivocally, to the rise of the factory system in cotton. Yet, in many other cases the distinction is less simple.

British firms were producing motor cars in 'mass' volumes by the 1930s; seemingly imitating the 'Fordist' system that had taken over the US auto industry by the 1920s. However, the detailed technology and practices of the British firms were qualitatively different from the

American industry, until at least the 1950s (Lewchuck 1986, p. 150). In these and other instances comprehensive change needed shifts in the mental framework of the relevant decision makers in the industry. The machines, techniques and tasks gell into a distinctively new set of production practices only when employers, managers and engineers act on a coherent new outlook – a paradigm – of how, and for what purposes, production is to be organised.

As defined in the sociology of scientific knowledge, a paradigm consists of a set of rules and beliefs defining the nature of scientific problems and the methods for investigating them. More recently, analysts of industrial change have referred to the paradigms shaping the pursuit of different types of manufacturing strategy and organisation. Perez refers to a 'techno-economic paradigm' (Perez 1985). Piore and Sabel talk of 'technological paradigms'. The first section of this chapter, tries to give these insights more precision. What distinguishes different forms of factory organisation is the production paradigms of their initiators and managers. A production paradigm does not specify the wider economic context of industrial production; being largely concerned with the relevant decision makers' product or sector. On the other hand, it is more than a technological paradigm. Within a production paradigm are forms of market conditions, administrative organisation, labour utilisation, and production methodology; plus specific product designs, machines and techniques, implied by the term technology.

To assess contemporary shifts in production organisation we need first to be sure of the scope and limits of preceding paradigms. In current debates technologies and paradigms are assessed according to the extent to which they continue or supersede Fordism and Taylorism. So, logically, the origins, character and compatibility of these two approaches ought to be clear. They are widely regarded, either singly or in tandem, as dominating manufacturing practice for much of the twentieth century. Closer examination, however, shows that both the range of phenomena to which they refer, and their mutual relationship is uncertain. Does Fordism simply mean mass production? Does it only apply to the production of consumer goods? Can it be synonymous with the entire system of Keynesian mass consumption capitalism? Is Taylorism therefore a minor part of the broader Fordist scenario? Or are commentators who identify one with the other misleadingly conflating two different regimes? Whatever the merits of labelling the mass production/mass consumption economic system as Fordism (Aglietta, Palloix, Sabel) here we will have to restrict the Fordism label to the industrial sphere, to manufacturing engineering and management. However, even in this more restricted sphere, Fordism, as a paradigm, can be seen to be quite distinct from Taylorism.

Industrial Fordism is here taken to be a methodology that reduces, simplifies and cheapens labour inputs through standardising both production operations and products – not necessarily in mass volumes – on the basis of the pre-specification of the dimensions of interchangeable parts. Its main organisational focus is the integration of the multiple operations into a process of production organised to achieve the fastest flow of work. Taylorism, on the other hand, will here be treated as a production paradigm which focused on simplification and routinisation of discrete tasks and work roles, irrespective of product characteristics. Its main focus is the hierarchy of administrative control over given aspects of the production process. The second and third sections of this chapter show the distinctions between these two paradigms by contrasting them to the hybrid workshop factory – an assemblage of related manual and machining operations, lacking fixed and systematic specialisation of functions. The impact which Taylorism and Fordism sought to have on workshop factory practice, described in later sections, shows their distinctiveness from each other. Using these yardsticks, later chapters clarify how far current developments are post-Fordist, post-Taylorist, or just continuations of these paradigms. First we need to establish some analytical and sociological reference points.

Some conceptual orientations

The factory distinguished itself from the workshop by organising production as a series of sequentially related operations; decomposing them into specialist operations, where practical and more productive. A centrally powered and automatically driven chain of machines assisted sequence and flow. It focused on the lateral, or horizontal dimension of production: the processing chain. A corresponding development for the vertical dimension of the same sites, was hierarchical administrative functions to control and command constituent sequences of operations, and co-ordinate them. Fordism – as a manufacturing rather than an economic paradigm – and Taylorism each arose primarily to deal with one of these two dimensions. Fordism focused on the process dimension to integrate and smooth its operations through standardisation of products and hence tasks. Taylorism, on the other hand, aimed primarily at establishing downward vertical controls over how particular functions were executed (see Figure 2.1).

Sociological and historical study confirms that although these two

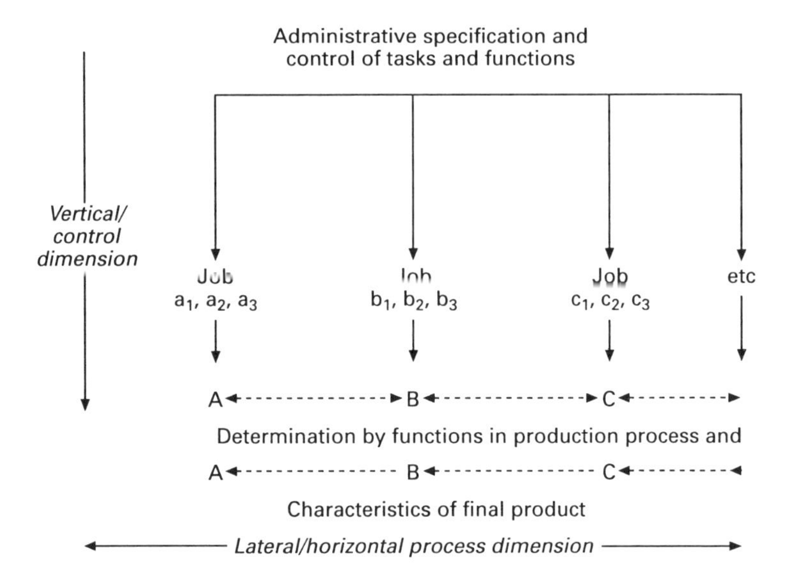

Figure 2.1 Taylorism, Fordism and the dimensions of factory production

major managerial movements of the twentieth century have become combined, their primary applications were not equivalent. The scientific management of Taylor and his supporters was a crusade aimed at factory and workshop managers. It was conducted, at least initially, by itinerant specialists and consultants on the fringes of top business management. Scientific management aspired above all to the same kind of professional status as engineers – from whose ranks Taylor himself had risen – and other 'servants of power' in industry. Taylorism arose as a response to turn-of-the-century problems; largely in batch-production manufacturers of capital goods. By 1920 scientific management as a special programme of change had largely been absorbed within US industry (Nelson 1980, p. 201).

The export of Taylorism from the USA received varying degrees of foreign acceptance; becoming largely institutionalised in various kinds of technocratic, and pseudo-academic societies. In some aspirant, or newly industrialised countries there was official support from state apparatuses. However, systematic application of specific Scientific Management techniques in European industry as a whole, came mainly after the Second World War. Its merits and pitfalls as a managerial package were largely played out amongst senior US managers after the 1920s. In Europe, as public interest in Taylorism waned, Fordism appeared as an alternative business system (Maier 1970, p. 54). Unlike Taylorism, the protagonists of Fordism were owners and top managers of large businesses, especially those selling consumer goods. It required strategies, and the authority, for replanning the relationship between the design and marketing of products, not just their production. Even in the operations sphere itself changes traversed all sequences of production in order to achieve mechanisation, transfer of work, assembly of interchangeable parts and speed of throughput. Ford's system crossed the Atlantic in the 1920s and 1930s. The scale and capital costs of implementation required action by governments and industrial tycoons rather than the engineering specialists and managerial consultants dispensing piecemeal, Taylorist solutions. Even so, as chapter 3 explains, adoption of both paradigms was more chequered in Europe. Widespread standardisation of methods, components and products, as well as the financial conditions for their mass consumption, had to await the combination of Keynesian economic policies with technical know-how from the US Marshall Plan after the Second World War.

The points to note here are as follows. Taylorism and Fordism were planned by different categories of economic actor. They had different spheres of application, and were disseminated at different periods. In short, we can infer that their scope and location in manufacturing operations differed one from the other. Yet each amounted to potentially revolutionary departures from the previous paradigm of batch-production workshops. Selected historical examples of this earlier paradigm help show the scale of their differences, both from this workshop system and from each other.

The world of the workshop

In 1835 Andrew Ure argued, in a vein later taken up by Marx (*Capital*, vol. I, p. 381), that the new type of factory was distinct from workshops since the former:

> designates the combined operation of many orders of work-people ... in tending with assiduous skill a system of productive machines continuously impelled by a central power ... a vast automaton, composed of various mechanical and intellectual organs, acting in an uninterrupted concert for the production of a common object, all of them being subordinated to a self-regulated moving force. (Ure 1967, pp. 13–14)

Notice how this model of the factory – as a 'system of productive machines', a 'vast automaton', 'subordinated to a self-regulating moving force' – anticipates many sentiments of the cybernation vision one-and-a-half centuries later.

The factory, according to Ure, comprised various textile mills and 'certain engineering works'; but excluded 'breweries, distilleries, as well as the workshops of carpenters, turners, coopers &c', because the latter lacked the central source of power and self-regulating machinery. However, it was not long before the spread of power technology made Ure's distinction between workshop and factory problematic. Central sources of power from steam engines were available for semi-automatic machines in most metalworking establishments by mid-century. Yet they retained more of a decentralised workshop, rather than a factory character. It is also now clear, as we will see later, that the factory division of labour extrapolated by Ure from the large textile plants – 'the combined operation of many orders of workpeople' – was much less pronounced in other sectors (Jacoby 1985, p. 14); especially in metalworking operations making small batches of products.

These operations, Ure correctly pointed out, subsumed the tasks of traditional trades into machines, rather than sub-dividing them along the lines of Adam Smith's pin-factory. In this respect Ure's definition of an industrial factory differs sharply from the Babbage definition – and that adopted by some modern theorists such as Braverman – of factory organisation as ever-finer divisions of detail tasks amongst more workers. Fusing together tasks into machinery is also more distinctive of Fordist practice (Williams et al. 1992, p. 524). The development of semi-automatic planing, key-groove cutting, and drilling machines superseded 'the dextrous hands of the filer and driller; while iron and brass turners are replaced by the self-acting slide-lathe' (Ure 1967, p. 21). Yet even in the early twentieth century all manner of machine tools in batch-production workshops still depended upon arcane and detailed skills to operate them effectively (Watson 1935).

The elimination of skill by automating batch production, faces the ubiquitous problem of the variety of output that is produced in small batches. Some batches are indeed large and may require simple metal removal operations. In these cases automatic or semi-automatic machines may allow the employment of operators with little skill or knowledge of metal-cutting technique. Their main tasks are to feed in metal blanks, and to stop and start the machine. Even here, however, the appropriate cutting tools must be set into the machine in specific positions or sequences at the start of every new job, and readjusted after tool wear or breakages. These are skills beyond the machine-tending operator, and so a division of labour between skilled setters, and semi-skilled operators occurs. The advantage of this occupational division is lost, however, when there are periodic switches from making large batches of the same product to smaller batches needing more frequent resetting. Then it may be as economical to retain a skilled worker as both setter and operator.

Such fluctuations in batches and their numbers may even make any use of specialised semi-automatic machines questionable. Automatic mechanisms render machines more complex than hand-operated manual equipment; and the first such machines could often only accommodate work of limited sizes and dimensions. Conversely, the hand-operated machines – where acquired manual skills were important – were more versatile in the range of work they could produce. Some workshops and factories would therefore judge the automatic machines a riskier investment than the more versatile manual ones (Brackenbury 1907, p. 220). When, as happened in Victorian Britain, the wages of skilled metalworkers was depressed by economic slumps – or by the victories of employers in pay disputes – then the cost advantage of investing in automatic machines would have to be weighed against the cheapened price of skilled labour (Staples 1987).

Most metalworking production in the early stages of capitalist industry retained a workshop character. With the exception of the more specialised makers of the new textile machines, each produced a range of diverse machines – steam engines, locomotives, distilling equipment, or machine tools – in small numbers for individual customers (Landes 1969, p. 183). This diversity in their products limited the extent to which machine and parts manufacturers could adopt special-purpose and semi-automated machines, based on innovations such as Maudsley's slide-rest lathe. For each new order would mean applying different machining techniques to make parts of widely varying sizes and shapes. Contemporary accounts suggest powered and semi-automatic devices were rare in most manufacturing operations in the early industrial period (Rolt 1965, p. 91; Smiles 1863, pp. 211–12). Maudsley's famous machinery for mass production of wooden blocks for ships turned out 160,000 units a year. Yet his small London workshop took six years to make the 44 machines required for this programme. Stephenson's, the dominant producer of locomotive engines, only made twenty-five engines in 1830–1 (Pollard 1965, p. 80).

The workshop of Roberts, later of Sharp, Roberts and Co. made the first automatic cotton spinning mule, and was mythologised by Marx – 'What is Vulcan compared with Roberts and Co . . . ?' (Marx and Engels 1986, p. 47). Yet in 1821 it consisted solely of hand-driven machine tools (Rolt 1965, p. 106). It was common for machine makers to build up their own equipment in increments. One semi-automatic machine, say a lathe, would be used to build the parts for another type of machine, and so on until there was a machine for every process. The Nasmyth and Gaskell works, which began like this, was still regarded as a general engineering factory – making machine tools and other industrial goods – even when expanding to supply the new railway industries. The products of Peel, Williams & Co – 'the biggest engineering business in Manchester during the first two or three decades of the nineteenth century' – included steam engines, locomotive engines, boilers, hydraulic presses, textile spinning machines and even gasometers. It never attempted the product specialisation conducive to the use of standard procedures and special-purpose machines (Musson 1960, pp. 8, 16–17). Whitworth's at Manchester, the pioneer large specialist plant for making machine tools, only emerged in the 1840s and 1850s (Rolt 1965, pp. 109–10, 121).

The Soho foundry – A failure of proto-Fordism

At least prior to the translation of mass production into mass consumer markets, most metalworking machine shops produced for capital goods. Their products were more likely to be required in small numbers. So the variety restriction on automatic machining continued to blur the technological and occupational distinctions between workshop and factory organisation. This constraint is sharply illustrated by the celebrated case of Boulton and Watt's early nineteenth-century Soho works. Several eminent commentators (Pollard 1965; Roll 1930; Wild 1975) have depicted this 'manufactory', established in 1795, as a precocious forerunner of scientific management and (Fordist) flow production methods. These accounts are partly influenced by futurist assumptions. That is to say that they over-emphasise the elements of contemporary factory organisation and management that can be discerned in what was essentially an extension of workshop practice. Roll's appellation of Soho as the 'first factory in the engineering industry of the world' (Roll 1930, p. 156) suggests a radical break with the workshop structures of general-purpose machines, varied products, and multi-skilled workers.

The new Soho works was managed by the sons of Boulton, the Birmingham manufacturer, and of Watt, who developed the steam engine made there. Roll equates it with twentieth-century establishments in four main respects. There was, firstly, a high level of integration of the constituent production process, which brought together the casting of the principal iron components with much of their machining and fitting on the same site. Secondly, a strategy of specialising in steam engine manufacture was carried out on the basis of a 'flow of work' organisation similar to contemporary mass production of standard products. Thirdly, elaborate procedures of cost and time controls were used to fix wage rates. Fourthly, and finally, a high level of occupational and task specialisation broke with the workshop tradition of the all-round skilled worker. At Soho, it is claimed, individual workers were assigned to one specific machining or assembling operation.

Roll equates the last two of these arrangements with scientific management of labour. The first two are described as analogues of Fordism; although Roll does not use this label. J. G. Smith's preface to Roll's study goes so far as to claim that 'Neither Taylor, Ford nor any other modern experts devised anything in the way of plan that cannot be discovered at Soho before 1805' (Roll 1930, p. xv). Such confident claims say more about the limited understanding of Fordist and scientific management systems in Britain prior to the Second World War, than they do about the actual status of production organisation at factories such as Boulton and Watt. A closer examination of organisation and methods at the Soho foundry shows that even the most advanced metalworking factories of the Industrial Revolution period were far from the integrated production processes of Fordism, and from Taylorism's hierarchic controls over work.

Take first of all the degree of integration of production processes at Soho. For much of the eighteenth century the new steam engines for mines, and subsequently for mills, were finished and put together in situ from semi-manufactured components, by teams of itinerant mechanics. Even the more standardised textile-making machines tended to be made to order, often by auxiliary businesses owned by individual mill owners. Boulton and Watt, and their sons, moved only partially away from this practice of localised construction. Their engines were still put together at the site of the mine or mill. Sometimes the parts were assembled together at Soho as a check on their compatibility; but key components continued to be purchased from outside contractors (Pollard 1965). Soho was therefore only partially integrated in terms of combining casting of the basic components with the subsequent major machining operations.

The second mooted characteristic of 'modern' production at Soho is the influence of a 'flow of work' purpose in the arrangement of the various machining and fitting sections (Roll 1930, pp. 186–7; Wild 1975). That is to say that the various machines and processes were located not according to similarity of their function – which became the dominant principle of the large workshop factory, the German *platzarbeit* system (Landes 1969, p. 306) – but according to the sequence of operations necessary to change blank metal into finished components and sub-assemblies. Yet Roll admits that the actual operations in these sections tended to be grouped according to the preponderant functions of the machinery: a heavy drilling (and boring) shop, a heavy turning shop (composed of lathes), a succession of fitting shops, and so on. Roll confusingly classifies the functional *platzarbeit* layout at Soho, turning in one section, drilling and boring operations in another, a kind of hybrid organisation, as a transitional stage between handicraft and mass production (Roll 1930, p. 187). Functional organisation is, however, the characteristic of the post-workshop factory. It was Bodmer's functional concentration of specialist machine tools in rows, according to the type of cutting operations they performed, rather than any flow of work principle, which became the dominant feature of the metalworking factory (Rolt 1965, p. 126). Functional layout has persisted in metalworking until recent attempts, often associated with so called 'just-in-time' methods, to organise on the basis of – sometimes computer automated – cells of diverse machines. The kind of flow production that Roll and others have tried to read into the Soho arrangements has to be recognised as prevented by two major obstacles.

Firstly, the range of components and products, and the metalworking operations upon them, makes it difficult to process jobs sequentially between successive work stations. Some single products or components could be made in a relatively straightforward fashion. Hypothetically, the production of large cylinders, for example, could be organised by first machining their end faces on a lathe. Then the actual cylinder could be bored on one or more suitable boring machines. Extra cutting of subsidiary apertures, 'de-burring' of rough edges and other imperfections, could then be completed in a final location. However, if cylinders of different sizes were required, these would need to be routed to different lathes or borers, or at least have the actual cutting tool changed at each work station. Moreover, to produce diverse components – pistons, gears or valves – as well as cylinders, then each machine section would need the capacity to work on a variety of quite different parts; either by separate machines or by machines possessing a range of different cutting tools. However, the different cutting operations for the different components would not necessarily be achieved in the same sequence – for example, the order of turning and boring might be reversed. So any particular section would have a range of different jobs entering and leaving; leading to potential problems in assigning priorities. In other words, steam engines, like most other metal machines, are not simple products that require only a few operations on a few components.

This point is related to the second major obstacle to flow production: volume. Where a high volume of the same products or components is required then premises and machinery can specialise in a few repetitive operations. If, on the other hand, volumes are low then it is clearly uneconomic to 'dedicate' specialised machinery or factories to single product/components. Soho suffered particularly from this barrier to mass, flow-line production. Boulton senior saw this connection from the start. He wrote to James Watt:

> my idea was to settle a manufactory . . . from which . . . we could serve all the world with engines of all sizes . . . with more excellent tools than would be worth any man's while procuring for one single engine. (Matthew Boulton to James Watt, 7 February 1769, cited in Roll 1930)

The volume of orders did justify the installation of new, special machines. However, the range of engines of all sizes, and the 'bespoke' nature of many of the orders – 'all the world', unfortunately, wanted their own special engine – prevented standardisation of the product and hence also completely special-purpose production machinery. As late as 1821 a visitor to Soho observed:

> 2 × 40 horsepower engines (HE), 2 × 20 HE, 2 × 10 HE, 2 × 40 HE for boats, 2 × 50 HE, several small engines for vessels of 20–15 &c HP. (Hall 1926, pp. 10–11)

Engines of six different sizes or functions were in simultaneous manufacture. The degree of machine specialisation that was achieved may have been assisted by the low levels of precision to which the machines made the components. Consistent with the practice of workshop production (Hounshell 1984, ch. 1; Best 1992) parts were machined to imprecise dimensions then corrected by subsequent manual operations – grinding, filing and tapping – to ensure a mutual fit. Thus it might have been possible to use the same machine for some similarly sized parts without overdue concern for precise differences in dimensions.

Systematic cost and time controls is the third contemporary manufacturing practice attributed to Soho in futurist accounts; because of Boulton and Watt's staff predetermining machine speeds and the job times for workers' piece-rates of pay. In 1801 optimal cutting speeds for several of the machine tools were calculated. It is presumed that these were then prescribed to the workers rather than letting them determine speeds. But this is not surprising as several of these machines were innovations, without customary standards of operation and not previously power-driven – Soho, naturally, used its own steam engines for power. More advanced, perhaps, was the calculation of the machining times for various components, in order to decide piece-rates. Did these practices amount to the Scientific Management of 100 years later? As the next section shows, Taylorist scientific management regarded job timing as incomplete without a prior task analysis to prescribe times and methods of work. Earlier versions did not apply these rates uniformly or individually to each worker. On several jobs the traditional system of 'internal contracting' (Littler 1983) survived. With internal contracting, a price for each job was negotiated between the employer – or employer's agent – and a foreman or master. The latter distributed the resulting earnings amongst 'his' work group at his own discretion. This was precisely the system that Taylorism attacked because it limited the direct influence of employers and managers over the individual worker. At Soho there was some internal contracting but no evidence of time-and-task analysis.

As for occupational specialisation by task, the fourth alleged characteristic of modern methods at Soho, it is true that this was greater than amongst traditional craftworkers in metal. Roll lists fitters, turners, and boring and drilling operators being assigned to tasks or machines specific to these functions, rather than doing a variety of jobs. But this arose, in part, because the artisanal millwright, or fitter using hand-tools, had not worked with specialist machines, and the corresponding higher degrees of accuracy. What did reduce the Soho turners' and millers' task variety was the relatively narrower product range and the required scale of machining. Fourteen different sizes of engines made for too much variety of components to allow standardising for a flow of work organisation. But this range was even wider in the contemporary mass of 'jobbing shops' – catering for the greater use of metal in diverse industries for all manner of machining and parts. Soho's lower variety in components and operations meant less opportunity to rotate machining, or fitting skills. The scale of several machining jobs meant lengthy operations to complete a single component – a total of 27.5 working days, including 20 days actual machining time, to make a 64-inch cylinder on Soho's large vertical boring mill (Rolt 1965, p. 71) – little chance then of such an operator rotating to diverse other tasks.

A more important qualification to the thesis of advanced task specialisation at Soho was that one form of specialisation was also an obstacle to 'modern' forms. Many of the 'specialist' machining occupations also included ancillary but substantial 'fitting' tasks. This is the main factor which restricted standardised work flows, limited specialisation and deskilling in workshop factories, of which Soho was the most developed. Despite improvements in measurement and accuracy there was still a great deal of uncertainty in the dimensions of the components to be assembled into sub-assemblies or machinery. The new lathes, etc. were capable only of accuracy up to about the width of 'a thin sixpence' or 'a thin sheet of paper'; in contrast to the thousandth of an inch developed through the American system of interchangeable parts in the 1880s (Rolt 1965, p. 187).

In the earlier factories to get all of the components to fit perfectly a variety of manual operations were performed subsequent to the machining process to achieve the 'dead fit' that became the hallmark of British engineering practice. But this meant – in direct opposition to the Fordist principle of accuracies predetermined by precise designs and interchangeable parts – that someone had to possess these very delicate fitting skills. At Soho, in addition to the specialist 'fitters', turners – i.e. lathe operators – also carried out such 'fitting' tasks as draw-filing and finishing, and fitting of gear wheels into sub-assemblies (Roll 1930, pp. 183–4). From my own inspection of the Boulton and Watt records it is also clear that the element of on-the-job specialisation was not, as might be expected, carried through into recruitment policies. There is no evidence showing that Boulton and Watt sought to employ, or develop workers with limited skills and experience. Lengthy apprenticeship arrangements continued and the crucial fitter skills were expected of most employees.[1]

Soho, then, was not an advanced precursor of Scientific Management and, Fordist, flow-line production, as Roll and others suggest. It continued to operate under the constraints of workshop practices: devolved controls and separate specialisations in craft-style work upon semi-autonomous operations, restricted integration of metal machining accuracy with assembly requirements, with wide variety in component range inhibiting flow between operations. Up to about 1850 the adoption of the semi-automatic machines, that Ure identified with factory principles, was slow. Until at least the 1830s most metalworking factories were still workshops in character (Pollard 1965, p. 80), with products still mixed enough to characterise them as 'jobbing shops' (Jefferys 1945, pp. 15, 53). The rarity of specialised machines and jobs was linked to the wide variety of products. The parts made were not easily assembled, identical, and interchangeable. These were manually fitted for final accuracy. Preferences for internal contracting of labour rather than direct management control, and the generally slow growth of a specialist management stratum inhibited administrative procedures to specialise and standardise those individual tasks that were repeated.

The first metal engineering factories of the industrial era remained largely 'workshop-factories'. Unlike the craft workshops of pre-industrial manufacture their products were more precise. They used some semi-automatic machines, and employed pattern makers, fitters, turners and planers, rather than millwrights and smiths with general skills. Yet most factories resembled more agglomerations of workshop units, than the integrated and hierarchically controlled factories of today. However, this analysis, of one of the most advanced 'factories' of the early industrialisation period, has provided a vivid illustration of how different these establishments were from full factory principles:

> it is a curious fact that the industrial mechanical progress of the last half century has not from that period marked any reliable principle of organisation by which one mechanical operation is distinguished from another. They seem to run into one another without any definite outline of distinction. (Fairbairn 1861, p. 28)

Contemporaneously, Marx was claiming in volume I of *Capital* that Modern Industry and its factory system had finally overthrown the Manufacturing phase of capitalism, because machines were now made by similar machines. Yet, even as he wrote, the majority of that production was still being conducted on workshop rather than factory principles.

Taylor and the control dimension

Contrary to a popular view, propagated by writers such as Braverman, Taylor was not the embodiment of systematic rationalisation. The spirit of Taylorism was detailed problem-solving. Taylor, a qualified engineer and ex-machine shop apprentice, was able to pioneer a highly specialised breakthrough in metal-cutting. By a punctilious development programme he increased machining speeds by improving the powering and metallurgy of the cutting tools. While this innovation stimulated a chain reaction in the development of more robust and powerful machine bodies (Rolt 1965, pp. 197–213), benefits accrued only gradually, leaving the overall organisation of machining unaffected. The explanation for the narrowness of Taylor's technical rationalisation is summed up in the title of his main text: not business management, nor even factory management, but 'Shop Management'– the organisation and control of the smallest individual unit of production within the enterprise.

The basis of Taylorism was that 'the foundation of the best management' is 'high wages and low labour cost' (Taylor 1947, p. 22). What was to be made, its design, and the organisation of the total production process were matters largely beyond the Taylorist frame of reference (Giedion 1948, p. 100). Some of his earlier successes were entirely confined to particular sections of the plant:

> his success, so clear at the shop level, had little impact on the overall operation of the firm, and left the economic benefits of his system in doubt. (Nelson 1980, p. 75)

The seventeen points in Taylor's 'scientific' manifesto suggest a preoccupation with piecemeal improvement of the status quo. Everything from the detailed running of an internal post and messenger service to an ordered system for classifying components is included in the seventeen points in *Shop Management*, partially repeated in *The Principles of Scientific Management*, published eight years later in 1911 (Taylor 1947). Yet only in one of these points, 'The Complete Analysis of All Orders For Machines or Work Taken By the Company' (Shop Management, p. 112, Taylor 1947), is there anything approaching an overview of the entire production process. Even there the integration or work flows between production functions is omitted. Taylor talks about 'standardisation'. By this, however, he means not the standardisation of parts and products to simplify operations but the replication of tools and methods when the same operation needed to be repeated subsequently (Shop Management, p. 123, Taylor 1947). While never stated explicitly the main organisational aspect of Taylor's approach is the bureaucratisation of management control. All of the seventeen points or functions, including the 'Pay Department', 'Employment Bureau' and a 'Mutual Accident Insurance Association', are to be the responsibility of a central Planning Department, which has the prerogative for analysing tasks, timing them, and fixing appropriate rates of pay; irrespective of the opinions of the shopfloor personnel themselves, or the shop supervisors. The latter become entrusted with executing Planning Department instructions, assisted by the ingenuously titled Shop Disciplinarian.

Taylor complained that his system had to be implemented in its entirety, yet this rarely happened in any firm that introduced scientific management methods (Taylor 1947, pp. xv, 212). In reality this is unsurprising. Versions of most of his methods had been tried out already, and the very detailed and composite nature of 'the system' allowed its constituent techniques to be taken up and used on a trial and error basis. Taylorism has a segmental character. Both the sociological profiles of the new 'scientific managers', and the industrial context of batch production created a preoccupation with the control and co-ordination of discrete operations, and an indifference to the constraints of product design and the flows of work.

Sociologically, as Taylor's own career testifies, the first generation of salaried and consulting engineers to tackle factory floor efficiency were subordinate employees or advisers to the owners and top managers of the emergent large corporations of the late nineteenth-century USA. They were hired to deal with pressing but essentially local production problems: not factory management but 'shop' management. As such they lacked both the power and perspective to attempt transformations of the factors structuring work in the individual shopfloor. The market qualities of the product, and the connection between product design and overall production processes defined Ford's paradigm, but were invisible to the Taylorists.

Industrially, because Taylorism arose directly from the batch production context of metal engineering, neither product standardisation nor systemic automation would have been plausible strategies. In contrast with Fordist principles Taylorism was limited in its scope. The essence of Fordism – at least as practised at the Ford factories – was not to correct 'inefficient' labour by studying and rationalising it, *à la* Taylorism. Instead, Ford aimed either to redesign the part or product involved, in order to minimise labour, or to substitute as much labour as possible by machinery. Sub-divided, short-cycle, single-task work roles were, in this sense, more directly effects of the available technology, rather than of the Ford strategy itself. So that in US car engine block production, most of these types of jobs were later eliminated by further pursuit of dedicated automation. By the 1950s, most of the remaining jobs were broader set-up and monitoring roles (Abernathy 1978, pp. 106–7).

By contrast, Taylorism oscillated between the technical details of given jobs and devising administrative controls over them. The technical focus arose from Taylor's own practical familiarity with the shopfloor and machine shop management. While advantageous in maintaining awareness of the details of production requirements, this experience also fostered a microscopic approach to factory problems. Organisations and procedures would be built up solely to analyse and control the detailed technical aspects of constituent operations. Thus Taylor's appointment at Bethlehem Steel began to be questioned because of the lengthy and costly experiments that he insisted on conducting into metal-cutting techniques (Nelson 1980, p. 89). Taylorists rationalising client companies became caught in a recurrent dilemma: the conflict between the firms' need for a quick and cash-saving solution and the Scientific Management credo: that administrative organisation for thorough technical and economic analysis and control must first be established. Of course, Taylorists knew machine operation, work organisation and shop management were interrelated and needed to be tackled together. But they also recognised corporate clients' expectations of rapid results, perceivable in dollar terms; rather than lengthy, detailed and rigorous experiments with the minutiae of operations.

At Bethlehem Steel, a Taylor company, there were eight types of official in the planning department by 1901. Their specialised responsibilities included not only costings, compiling instruction cards for specific jobs, time study, and production resource needs, but also the balance of stores, the order of work, and its routeing through the shop. Between forty and fifty 'functional foremen' then had the necessary powers for imposing and maintaining the resulting plans. Most of these specialist functions have appeared in metalworking plants since Taylor's day. Their limitation was the sudden scale and centrality of such an administration, and the associated relocation of authority within the plant. Taylor's bureaucratic blueprint was rarely adopted with the level of complexity that he proposed. Even at the Link-Belt Company, one of the most successful of Taylor's projects, company managers later reduced the numbers of planning department staff by 40 per cent. In 1910 a US government inspector reported that Bethlehem – a major arms contractor – had too much detail in its planning department and that the numbers of clerks and foremen should be cut to bring down costs. Taylor's reforms at Bethlehem Steel took up more than three years. However, when he was sacked in April 1901, he had still not worked out, let alone applied, his famous differential piece-rate system in the company's crucial Number 2 Machine Shop. Hence these time requirements meant a constant temptation to by-pass, or suspend, the investigation of organisation and methods and apply the quick fix of a time study and piece-rate scheme, wherever in the plant it seemed likely to bring quick results. The latter option encouraged the impression that isolated changes were feasible, thus further undermining the systemic aspect of Taylorism. As chapter 3 will explain, this slippage had profound consequences for the spread of Taylorism in Europe after the Second World War.

Taylor's central aim for work roles was always to simplify the range of discretion and task complexity required of individual operators. In machine shops this invariably meant diverting attention between technical improvements on individual machines and administrative controls over their operators. Thus tool making and sharpening were segregated into a separate department; and specialist supervisors ('functional foremen') were given responsibility for calculating and prescribing optimum machine speeds for each operator. Taylor knew from his own experience the difficulty of devising general rules and procedures for machine tools, because most differ in uses and functions. Yet the commitment to 'science' meant universal laws. Taylor's own protracted attempts to resolve this paradox only furthered the contradiction between 'science' and profitability in metalworking batch production.

Taylor's technological contribution of high-speed metal-cutting, was coupled with optimum maintenance procedures for the belting which transmitted the power of steam engines to each machine. These were universal laws of machine use. For one justification for the metal cutting experiments was to replace particularistic judgements and practices on individual machine tools by a general set of formulae – a 'science of metal cutting'. Hence the strict Taylorist rubric that task simplification and operating standards must be preceded by the micro-analysis of the capability and limitations of individual machine tools. But the particularities of batch metalworking served to restrict further the benefits of Taylor's innovations. By the turn of the century, electric motors were powering individual machine tools; rendering superfluous all the experimentation with belting – as Taylor himself came to realise when he adopted an electric motor for the last of his cutting tool experiments (Rolt 1965, pp. 202–3). Even his new cutting technology was of little practical use at first, because speed-changing devices and electric motors had to be developed for the high-speed tools.[2] As with the first computerised machines in the 1960s, these requirements held up the adoption of high-speed cutters, since only 'large and progressive firms could afford expensive alterations to their machine shop operations'(Nelson 1980, p. 85).

An inherent economic flaw in the manufacture of metal goods is time wasted between one operation and the next. Fordism later succeeded, in some sectors, in minimising such costs by flow methods and standardising the product. Characteristically, however, Taylor's solution – as instanced in his encounter with the problem in the efficiency drive at Bethlehem Steel's machine shops – required more effort from and more administrative control over, the human counterpart to the high-speed cutting tools: the individual operator. The effort problem was defined as a problem of motivation, solvable by bonus and incentive payments. As chapter 3 will show, the predominance of this approach in British rationalisation after the Second World War, created further social impediments to Scientific Management.

Taylorism lacked the Fordist insight that simplifying the range of parts and operations into fewer, more standardised products could enhance work flows. Instead, an omniscient planning department was to exercise hierarchic control of the existing flow of work. But the accompanying functional specialisation meant a multiplicity of planning staff. Taylorism thus meant both a base of incremental time study and piecework schemes, and a potentially complex bureaucratic superstructure. But the factories struggling to evolve from the workshop still combined a variety of products and operations, which needed more systematic solutions than central control of multiple tasks and times. As the British reception to scientific management discussed below shows. Taylorism's one-sided emphasis on the vertical dimension of admininistrative planning and control entailed further internal contradictions. In its full-blown form it also presents an acute contrast with the Fordist paradigm.

Fordism: process integration through standardisation

Where Taylorism placed special emphasis on pressure from vertical bureaucratic controls, Ford aimed at minimising such measures. Classic Fordism also meant: equipment dedicated to a single, standardised product; components with invariant and pre-specified tolerances; and a labour relations policy that made even fewer concessions to worker autonomy than Taylor's simplistic model of the income-maximising, work-shy, but negotiable employee. With the partial exception of tool makers, Fordism came to view production workers as ready-made, and readily expendable, mechanisms for insertion into pre-defined places and functions in the overall process of production. These principles of Fordist work organisation and labour relations, but not always the eventual practice, follow from its manufacturing methodology.

Fordism has industrial relations, political, and economic regulation aspects, but as a manufacturing methodology it differs from other production paradigms in its single-minded focus on the lateral, process, dimension of production. It has been argued (Sward 1948, pp. 33–4) that Taylor personally influenced the adoption of flow-line organisation for machining sections at Ford's Highland Park plant from 1910 onwards. This point is debatable in view of Ford's own disavowal of the influence of scientific management.[3] We have already seen that changing production flows themselves was not of central concern to Taylor. But even if this were to be the case the organisation of production on a flow-line, rather than functional basis is a derivative rather than a primary feature of Ford's breakthrough.

The inner logic of Fordism, as a manufacturing system, is not the moving assembly line. Nor is it the fragmentation of production jobs, nor the mass production/mass sales equation; both of which had been practised well before Ford's innovations. The moving assembly line was simply a technical fix that perfected the pre-existent manufacturing logic. Moving assembly, fragmented production jobs and mass production/sales arose out of the pursuit of a standardised product made from interchangeable parts. A single standard product whose features could be specified down to the finest detail and dimension needed mass production and sales to justify the attendant specialisation of capital equipment. These requirements, in turn, called for maximising the integration of each phase of manufacture; potentially from the original raw material to the finished commodity.

Ford's River Rouge plant is an extreme example of such process integration. The Model T automobile – the paragon of the standardised product – was already eleven years old when the new River Rouge complex began making it in 1919. Spread over 2,000 acres, 'the Rouge' had its own deep-water port, railway yards, power station and steel furnaces 'upstream' of the machine shops and assembly lines for the Model T and trucks.

Single-product specialisation, of the Model T, lasted from 1908 to 1927. It allowed Highland Park, and then River Rouge to use single-purpose machines, dedicated to continuous production of the same part, to the extent that the machining of engine cylinders could be decomposed into twenty-eight separate operations, each performed on specialised types of machine tools dedicated to that operation and no other (Arnold and Faurote 1915, pp. 73–7). Significant resetting of fixtures and tools was minimised; as was the risk of variations in the specified dimensions of the components and operations. Ford managers claimed that even 'green' labour could be trained up in a couple of days to undertake such specialist jobs (Arnold and Faurote 1915, p. 41). Dedicated machines could provide a constant supply of a limited range of identical components which could be assembled by specialised workers; without Taylorist task analysis. Operators worked as interchangeable members of particular section teams (Arnold and Faurote 1915, p. 29), uncannily like contemporary Japanese just-in-time arrangements. Machine tool dedication preceded and made possible the automation of the transfer of parts between work stations, as well as the eventual installation of moving assembly lines. However, the basic feature distinguishing integrated production from the most distinctive previous paradigm – British-style precise manual fitting between machining and assembly – is the principle of interchangeable parts.

Before Ford, American makers of domestic and capital goods – such as McCormick with harvesters and the Singer sewing machine – had adapted the interchangeable parts system from its origins in US arms manufacture. Under such 'armory practice' (Hounshell 1984) accuracy of final fit was achieved, not by manual filing and adjustment prior to assembly, but by prior calculation and specification of a tolerance range outside of which machining must not vary. Precision control was assured by the provision of precise templates, in the form of jigs and fixtures. At each operation, prior to assembly, dimensions were checked for conformity to prescribed tolerances by special gauges. So at the final stage of putting components and sub-assemblies together, conventional 'fitting' work was entirely superseded by simple assembly; which required no additional adjustments since the accuracy of the components had been pre-determined and guaranteed.

It is this principle that further segregated metal removal skills into the 'upstream' machining jobs, and simplified tasks and skills in the downstream assembly of the product. It also helped to systematise the flow of work; and thereby reduced the time and tasks required of assembly operators. The jobs of the machinists, on the other hand, were simplified more by standardised products than by the interchangeability of parts. Later chapters will show that it is only with contemporary computerisation that the interchangeability principle is being applied to the machining stage itself. The pre-calculation of tolerances is combined with pre-determination of methods in computer programs to simplify the machining stage. Just as machining to specified tolerances simplified final assembly functions 100 years ago.

Ford's own practices are relevant because, like Taylor's experiments, they constitute limit cases of the respective paradigm. They show the most clear-cut characteristics. Ford and classical Fordism soon diverged. The Ford Motor Company was forced to drop first its single product approach – as far as automobile production was concerned, Model T Fordism became exceptional (Hounshell 1984, pp. 300–1) – and then its labour relations practices. The labour relations aspect of classic Fordism required certain favourable social conditions. In its early days Highland Park had a ready supply of un-unionised labour for the semi-skilled jobs mechanisation had created. With the moving assembly line, and overhead conveyance along flow lines in the machine shops, workers' pace and performance were dictated by management's control of line speeds and the ease of identifying ineffective work. This technology also enabled Ford to replace incentive payments with flat-rate pay (Sward 1948, p. 48). But standardisation and speed-up also bred individual volatility and threats of unionisation to challenge managerial authority. In one month in 1913 Ford was losing, or having to sack, half of its total production workers. Most of these left voluntarily; because of the harsh work practices, or for higher wages elsewhere, or for both these reasons. Three times the numbers needed were being hired simply to retain the third who would stay longer.

Simultaneously, and again in 1919, industrial unions tried to organise Ford's and other firms' auto industry plants in and around Detroit (Meyer 1981, pp. 90–4, 171–94). Ford tried first to fight the union threat with the process integration principle. In 1912, and again in 1919, he bought up component and body-plant supply firms affected by disputes and strikes. Their operations were transferred inside Highland Park and River Rouge

(Meyer 1981, pp. 91, 193). His second response was to award more responsible, and longer-serving, employees higher rates of pay: the legendary 'Five Dollar Day'. Unlike Taylorism, Ford had anyway always refused output-related pay (Meyer 1981; Ford 1924, p. 126). The third response was to begin changing labour–management arrangements, towards more Taylorist forms. Out went arbitrary powers of the shop foremen; in came specialist personnel managers and policies. The Five Dollar Day system brought the first time-study of jobs. The status of labour in Fordism was transformed. As one journalist put it: 'the human element' at Ford changed from a 'variable' factor to one that was 'constant'. From the late 1920s the market conditions of 'pure Fordism' also deteriorated. Complete standardisation and integrated production became more complicated as other producers competed on the basis of model variety. Although Ford continually sought change in production methods, components and equipment, in order to lower costs (Williams et al. 1992), the Model T itself remained adamantly unaltered. However, General Motors developed its distinctive market strategy of segmenting the market with different models for different income groups, and annual changes to these models. The US motor car industry moved into what has been called 'flexible mass production' (Tolliday and Zeitlin 1987). Ford lost sales and had to move away from both the Model T and completely standardised mass production.

The labour problem and the new operating constraints of flexible mass production interacted to modify 'pure Fordism' further. Deskilling, although not reversed, was at least halted. Annual model changes, though often cosmetic, limited the lifetime of specialised machines. Redesigns of engines, bodies, and drive systems meant periodic retooling and running-in of equipment operations. On model change-overs the adaptability and know-how of experienced production workers could reduce the speed at which the change-overs were accomplished. On the machining sections additional, complex retooling, maintenance and die-fitting tasks renewed the importance of mechanical skills. As described above, more automated machine tools, combined with more model changes, eliminated repetitive manual tasks and led to higher task responsibilities for the main machining work on engines.[4]

The 'variability' of labour under Fordist automobile production was further checked from the mid-1930s when Ford finally had to accede to unionisation. Thereafter the burgeoning personnel bureaucracy forever lost its original leanness, in order to administer and regulate the complexities of what became the legal-contractual system of US industrial relations. From this system a job classification complex – more fully analysed in chapter 5 – subsequently evolved. This brought a measure of co-regulation into task definitions and the movement of workers into, within, and out of the company. The original Fordist treatment of labour as a homogeneous, malleable raw material was thus decisively diminished. Despite these modifications the primary thrust of Fordist standardisation took the auto industry off on a separate trajectory to the normal metalworking batch producers. Their attempts to emulate the automobile sector are the subject of chapters 6 and 7. For the moment we need to backtrack slightly to take stock of mainstream metalworking's initial responses to Taylorist and Fordist innovations.

The workshop factory meets Taylor and Ford

Other firms changing their production methods and organisation did not, of course, see this as a choice between Fordism or Taylorism. They saw a range of strategies, methods and techniques from which they were free to select the practices best for their own operations. Implementation often meant noticeable overlaps between Taylorist and Fordist approaches. At the Ford Motor Company itself, hierarchic controls, such as the time-study techniques for individual jobs, were tried out during completion of the new assembly methods (Sward 1948, p. 47). Ford's extensive deployment of overseeing supervisors was a substantial, though far cruder method of hierarchical control, than the Taylorist variety. Analogously, the Tayloristic promotion of more accessible tools and workpieces approximates Ford's use of conveyors and assembly lines to minimise the task motions of his line workers. Yet most of these similarities are outweighed by the massive differences in their main targets for change. Taylor's emphasis was upon a bureaucracy of specialists with methods, resources and bonus calculations. At the Ford Motor Company, on the other hand, administration was minimised. In expulsions rivalling the Party purges in the USSR, Ford cut 75 per cent of supervisors, 50 per cent of the office staff, and the entire statistical department in 1921 (Sward 1948, pp. 79–80). Moreover, the individual worker was there to be driven, rather than organised and motivated. In tasks simplified by the combination of mass production and standardised – interchangeable – parts and methods Ford had developed an apparently perfect driving force that made Taylor's elaborate and painstaking procedures seem pedestrian. At key points, such as the introduction of the $5 and $6 day rates, Ford's command of the integrated technology of production allowed major and arbitrary speed-ups (Sward 1948, pp. 77–8) that would have been anathema to Taylorism.

As complete solutions to variability in production and human labour both approaches were weakened by their concentration on one axis of production organisation. To routinise production Ford depended on a single standard product and hence highly specific competitive and market conditions. A full-blown Taylorism demanded a synoptic and systematic bureaucracy to replace existing business administrations; an even more elusive, almost impractical, condition. The new paradigms did penetrate the batch production metalworking factories, but in geographically and organisationally uneven ways. In the USA these plants showed – in addition to the piecemeal adoption of scientific management already mentioned – an early interest in some key Fordist principles. Indeed, industries such as sewing machines, clocks and bicycles, adopted interchangeable parts before, or in the same period as Ford systematised them (Layton 1974, p. 380). Product specialisation and interchangeable parts, plus a mass production–mass consumer strategy, would allow something like Fordism. Elsewhere in US industry changes were more eclectic. Machine tools, a classic batch-production sector, rapidly took to interchangeability and some product specialisation but, without the full paradigm, it became stuck between Fordism and Taylorism.

The US machine-tool industry

This sector seems to have adopted mechanised production equipment later than the British. The Ohio craft-engineer Hartness had to fight hard to turn Jones and Lamson into a specialist lathe manufacturer in the latter half of the nineteenth century. He reported that, as late as the 1870s, most workshop-factories used hand-tools, and only three types of mechanised machine-tool. British-style hand fitting to a maximum accuracy of a 64th of an inch prevailed (Broehl 1959, pp. 12–13). Firms dedicated solely to machine-tool products emerged from the 1880s onwards, partly to supply the new mass production industries. Then the rapid growth of larger markets fostered a dual structure (Herrigel 1991); some firms concentrated on making standard kinds of specialist machines, such as turret lathes and screw cutting machines, the rest made a range of varied machines more customised to individual buyers' preferences. Until demand shifted to more complex machining systems from the 1930s onwards, the dominant trend was making standardised single-function machines for large, although not mass, markets.

With this quasi-Fordist product specialism came attempts to minimise skill and task variety in the production of standard special-function machines. In these firms special jigs and fixtures limited the range of expertise required to operate machine-tools. Taylorist central planning sought to homogenise machining sections and simplify skill repertoires for each type of metal-cutting operation involved. Yet the limited scale of demand for machine-tools still precluded the mass output of pure Fordism. Owners' and managers' production paradigms resisted large, integrated, single-product operations. In the First World War the whole industry rejected government attempts to rationalise production into fewer standard products.

After this war, paradoxically, the automobile industry itself inhibited further progress towards a Fordist market structure. The expanding car firms became dominant customers for machine-tools but they wanted special-purpose machines unsuitable for other sectors. The machine-tool industry compromised by making and using standardised components for different types of machine product. Even state-directed production specialisation in the Second World War hit natural constraints to mass production techniques: the high precision criteria limited the application of low skill and broad tolerances in machining and assembling (Broehl 1959, pp. 96, 193). As chapter 3 will show, the scope for further standardised production was still an issue after the Second World War. Unlike Britain, however, expansion of these industries in the USA seems to have meant relatively rapid falls in the ratios of all-round skilled workers on machining work (Jerome 1934, pp. 295, 400); perhaps because making special-function machines needed less skill (Herrigel 1991, p. 239).

Britain

Responses to Fordist and Taylorist techniques in British batch producers were even more eclectic and particularistic. In Europe generally it was a second-hand Taylorism that was disseminated through publications and 'experts' who had little or no direct experience of how scientific management was practised in the USA (Devinat 1927; Layton 1974). In Britain, managements were already experimenting with various incentive-payment schemes such as the Rowan and Premium Plan systems (Taylor 1947, pp. 42–3). So Taylorism, at least until the Second World War, was absorbed into a paradigm of work role improvement rather than systematic rationalisation of planning and administrative controls. Acceptance of the even more transformatory Fordist paradigm was limited. In the first place this was because managers and engineers doubted the utility of the interchangeable parts system. Later the older preoccupation with individual work roles, and the market limitations on mass production, dominated perceptions, even amongst automobile manufacturers (Lewchuck 1986).[5]

When scientific management and Fordism were being adopted in the USA British engineers and managers recoiled from the political and financial costs of administrative reorganisation which they detected in the new ideas. To the practising British engineering manager Fordist and Taylorist prescriptions were indistinguishable aspects of an alien new wave of untried innovations (cf. *Transactions* of the Institution of Engineers and Shipbuilders, 18 December 1917). Some of the responses to the prospect of standardised parts and products were partly ideological. It was argued that their adoption could suppress initiative and innovation in the workforce (*The Engineer*, 25 January 1907, p. 93). But such psychologistic caveats did reflect the persistence of unionised job controls and traditions of skilled workers which were much more deeply entrenched than in the USA. Market uncertainties also loomed large. How would standard products be received by clients long used to making and receiving special orders and special product features (*NE Coast Institution of Shipbuilders and Engineers* [1917] p. 116)? Both scientific management and Fordist policies threatened more expenditure in administrative overheads.

Standardisation meant greater preparation between the different functional stages of design, costing, provision of tooling and materials, and the actual production processes. Costs of more specialised non-production staff were bound to rise to get these higher levels of co-ordination. Advocates of Fordism could argue that higher overhead costs would be spread over the higher volumes of output if specialisation and standardisation were also achieved. Others, however, objected that either the technical character of their product, or the resistance of financially conservative boards of directors made such strategies impractical. In addition, the general fixation with direct worker productivity could always be invoked. Why go to all the trouble of product, market and organisational transformation when the improvement of worker motivation and commitment constituted a simpler way to higher output and lower costs (Martin 1907)? Even heads of potential mass production car firms, such as Herbert Austin, saw individual worker responsiveness as central. In an astounding play on words, another car firm boss, F. W. Lanchester – adapter of interchangeability to British car production – inverted the concept of a co-ordinated flow of work into an issue of worker commitment:

> If the men urge the staff, the works organisation flows like a river, whereas if the staff urged the men, it is more like a pumping system in which you have pumping going on at regular intervals. (Lanchester 1907, p. 132)

British bosses were Tayloristic by instinct, even in latently Fordist trades. In Europe Taylorism's profuse presence in publications and quasi-professional institutions gave it a head start on Fordism. The range of Taylorist practices and its ideological framework made it a more convenient store for selecting specific and discrete solutions. Even as late as the 1950s although Woodward's major survey of organisations in Essex found only one firm with management by pure Taylorist functional specialisation, elements of Taylorist production control were common in other firms (Woodward 1980). Fordist practice did lead firms to set technically achievable yardsticks for production techniques and organisational principles. In the early twentieth century, skilled fitting declined with partial standardisation, through pre-specification of component dimensions by more systematic jigs and fixtures (Zeitlin 1987). Demonstrations of the time savings from regrouping machines according to the sequence of operations – 'flow of work' – for particular products challenged the prevailing functional deployment of machine-tools for the first time. But many British markets were either narrow or unsuited to mass production and marketing; or firms lacked the will to promote standardised products in them. So, a piecemeal Taylorism generally had a more immediate appeal to British producers than Fordism; albeit limited by older traditions – especially employee relations and the shopfloor regulation of work.

Conclusion

Contrary to observers such as Ure and Marx, the rapid, early nineteenth-century takeover of integrated factory principles, in such leading industries as textiles, was not amenable to other sectors of the Industrial Revolution, such as metalworking engineering. Factory techniques were adopted there, but partly because the traditional artisans and workshops had never made the new products – mainly the precision components for industrial machinery. The new equipment developed to make these involved discrete and partial automation of individual machines, such as lathes and planers to meet particular needs.

Thus developed the hybrid 'workshop factory' with its mix of some mechanised machining operations and more laborious manual and semi-mechanised processes. Three factors impeded machinising: product variety, filing and fitting of parts manually for an accurate fit, and the need for administrative organisation to control the flow and execution of work. Inability to market a specialised product, to develop systemising to uniform dimensions, and reliance on autonomous workers' skills – often organised by 'internal' contracting – limited pioneering attempts like the Soho works. This distinctive British manufacturing paradigm was unlikely to transform itself. Change came instead from the different conditions of North American industry. There the development of interchangeable parts – initially by small-arms contractors to the US government – began to eliminate the need for skilled fitting work and constrained the independence of the machinist. In addition, the larger sizes of potential markets – together with interchangeable parts and relatively greater shortages of skilled labour – favoured specialised products. Ford brilliantly and ruthlessly exploited both of these developments to integrate the discrete processes for manufacturing small and medium batches and to decompose and mechanise what had previously been skilled work practices. Ford perfected what other firms had already tried. However, because the systematic organisation for making the Model T stemmed overwhelmingly from the integration of processes and the standardised and simplified product, Fordism seems a condign label for this paradigm.

The administrative job-by-job controls pursued by Taylorism were, at the most, secondary factors in the Fordist approach. The main concern of the latter was, primarily, integrating the constituent manufacturing processes on the basis of a specialised product made to standard and rigidly determined dimensions. Taylorism differs because its focus is to rationalise and standardise the tasks for discrete and individual work roles through hierarchical administrative controls. For Taylorism, process integration is secondary and pursued mainly through the elaborate bureaucracy of the planning function. These then are the paradigms. But the devil is in the detail. So far we have only examined their realisation in the USA. Were Taylorism and Fordism adopted and practised uniformly in other countries? How have they actually been practised in the complex environments of specific sectors in other countries? We have seen how paradigm conflict initially checked an early transfer of Fordism and Taylorism into British industry. Yet some aspects have come to be widely adopted. How? How far? With what compromises has this diffusion occurred? The next chapter shows how, through the agency of the Marshall Plan after the Second World War, a wider range of Taylorist policies – specialised management functions, machine utilisation, time study and systematic costing – became widely adopted (Woodward 1980; Carter 1976). A comparison between Britain, Italy and the USA shows, however, how historical time lags and national institutional contexts, have checked the possibility of a uniform march of Taylorism and Fordism down their chosen axes of hierarchic control, and standardised and integrated products and processes.

3 Productivity for prosperity: industrial renewal and Cold War politics

By the onset of the Second World War, Taylorism and Fordism were successfully entrenched in many US workplaces. Successful alignment of a modified mass production paradigm with mass markets, by Ford and other manufacturers, had established a yardstick for other batch producers in metalworking. This course was facilitated in some sectors by the large markets that existed in the USA for many industrial and consumer goods. Assisted by the rapid professionalisation of engineering managers – illustrated by the emergence of 'industrial engineering' as a specialism – many Taylorist management methods also became standard practice.

The relative weakness of US trade unions in this period and the responsiveness of the new paradigms to economic conditions might suggest the latter were the outcome of a spontaneous logic of organisational and technological development. However, the sharp contrast in the application of these paradigms in the European economies shows how a specific technological trajectory is affected by the relatively autonomous social and political trajectories. The extent to which British and Italian industry took up Taylorist and Fordist approaches is largely explicable in terms of the international politics of the post-Second World War period, and the domestic politics of labour relations in the same era.

This chapter aims to show the scale of the gap between the USA, with its Tayloristic controls over labour, and attempts at Fordist standardisation and simplification of products and processes, and the indigenous eclectic practices that had evolved in Britain and Italy. Contrary to the chronologies of some writers (such as Littler 1982, pp. 144, 188), we will see that the two decades following the Second World War involved the most systematic attempts to transfuse Taylorism and Fordism into these countries. However, the adoption of these paradigms was differentially modified in each country by pre-existing social institutions, and by interaction with simultaneously developing socio-political trajectories. Moreover, practitioners and policy makers, at that time, saw themselves as trying to transfer as much as possible from a common stock of 'American' management and production techniques. Few participants made distinctions between the Fordist and Taylorist elements in this repertoire.

The general aim was to secure adoption of those techniques that would enhance productivity. From the perspective of the emerging consensus on pro-Fordist economics, productivity was the key to raising general living standards to the level of the USA. West Europe would then be economically and culturally insulated from the rival political challenge of USSR communism. On these assumptions the priority was to adopt those North American techniques that would achieve rapid and generalisable increases in productivity. More basic product and market factors that restricted the batch production factory to workshop forms of organisation were not so readily appreciated. In both Italy and Britain US economic and political diplomacy propagated alliances of state technocrats and enterprise managers. However, these opted for a modified Taylorism as the solution to productivity deficiencies; although the form of Taylorism differed according to national differences in the balance and configuration of political power in labour relations.

Broadly speaking the political split between left and centrist unions in Italy, together with the dominance of a US-backed anti-communist coalition government, encouraged enterprise managers to use Taylorist techniques in a strategy of sweated labour. The simplicity of enhanced profitability by this route limited interest in the more sweeping product and process technology changes of Fordism. In Britain, on the other hand, the influence gained by shopfloor unionism during wartime productivity drives, was augmented by the participation of national union officials in quasi-corporatist policy making; of which industrial rationalisation was one strand. Union influence did not reach its height until the 1960s, but the need for a measure of consensus led to a Taylorism based primarily on *negotiated* payment-by-results schemes.

This diluted form of Taylorism allowed the maintenance of craft-style working practices. Moreover, management reluctance to adopt Fordist rationalisation of organisation and technology, partly because of market constraints, limited the standardisation of products and processes. European batch production after the Second World War thus concentrated on Taylorist hybrids, aware of Fordism's greater potential but largely incapable of realising it.

To demonstrate the detailed development of these patterns involves examination of the institutions of the technical assistance programmes of the US Marshall Plan. We will then need to consider the central images of US practice and institutions which influenced British Marshall Plan participants, before going on to examine how – and to what extent – unions and managers, first in Britain and then in Italy, participated in the trans-Atlantic transfer of ideas and techniques.

The Marshall Plan and the old world's view of the new

Left to themselves many European manufacturers would probably have been unaware or sceptical of the techniques and organisation used in the plants of their US counterparts. However, in the immediate post-war years Western governments and policy makers saw Europe's conversion to these techniques as a political necessity. Manufacturing modernisation was a central feature of the plan initiated by US Secretary of State, George C. Marshall in mid-1947. Through a variety of policies, the Marshall Plan sought to create flourishing economic foundations for Western European politics. Generalised economic prosperity was to help to insulate European populations from communist influence (Hogan 1987). Financial assistance was provided to correct immediate monetary crises. However, for Marshall planners reconstructing these economies also came to mean policies for new industrial practices: 'gains in specialisation, economies of scale, and labour productivity ... to drive down prices, make their commodities more competitive, and balance their accounts with the dollar area' (Hogan 1987, p. 49).

The means for industrial reform were threefold. Firstly, information and assistance were disseminated by the US Economic Cooperation Administration through establishments such as the Anglo-American Council on Productivity (AACP) and the Italian National Productivity Council. These were to spread awareness of the scientific, technical and organisational bases of American industrial success. Secondly, through the participation of US union leaders, non-communist trade unions in Western Europe were to be split off, organisationally, from their communist counterparts. The independent unions were also to be won over to the principles of North American styles of co-operative bargaining based on sharing of productivity gains. Finally, the scope for economies of scale was to be boosted by creating a supra-national West European trading bloc, that would transcend the existing national restrictions on market demand and, by implication, on mass manufacturing.

This last aim – the most crucial for establishing full-blooded Fordist mass manufacturing – largely failed because the British government resisted economic union; though the idea can be seen as contributing, indirectly, first to the founding of the European Economic Community and, belatedly, to the post-1992 abolition of all internal trade barriers in the Single European Market (Cecchini et al. 1988). Meanwhile, in Italy government and business leaders rejected the higher wages necessary to expand consumer demand at home (Carew 1987, pp. 162–3). The second aim – of industrial relations practices more favourable to business expansion – figured in the withdrawal of both US and European union federations from the communist-aligned World Federation of Trade Unions.

This schism isolated the communist-led labour unions in countries such as France and Italy, but did not easily convert others to a putative US-style 'business unionism', of co-operative productivity bargaining. British unions' support for movement away from job controls that business saw as 'restrictive practices' was qualified and ambivalent (Hogan 1987; Tewson 1953). Moreover, as the following account shows, the US image of co-operative, productivity-oriented unionism was often not a reality. However, the first above-mentioned Plan policy – spreading knowledge and acceptance of manufacturing ideas and techniques – seems to have had some success, even though the transplanted technical 'know-how' had a narrower focus than in the US.

The AACP teams who made the pilgrimage to the USA in the 1940s and 1950s formed a concerted flow of trans-Atlantic influence into Europe. They served as models for the measures later recommended and adopted on a more modest scale by several other European nations, although the corresponding inflow to Italy was weaker and more uneven (Carew 1987, pp. 162, 176–7). AACP reports are significant for understanding both the precise methods and approaches disseminated, and the state of Tayloristic and Fordist practice in the USA. The AACP first met in London in October 1948. It identified five priority areas in which British industry should adopt North American practices: transfer of production techniques; maintenance and improvement of productive plant and power; productivity measurement; economic information; and a triple 'S' formula – standardisation, simplification and specialisation in production (Anglo-American Council on Productivity – 1952, pp. 6–7, 12–13).

Information for these topics was gathered from forty-seven different sectors, and seventeen 'specialist' groups studied topics ranging from 'education for management' to the techniques of hot-dip galvanising (AACP 1952, pp. 9–10). Sectoral study teams tended to reflect the priorities of the Council's committees: notably production techniques, equipment and the triple 'S' formula. The teams were 'advised to prepare a detailed questionnaire covering all aspects of its terms of reference, e.g. technical processes, types of plant and equipment in use, handling methods, materials and components, production control, costing, industrial relations etc.' (AACP 1952, p. 20).

However, teams tended to interpret 'industrial relations' broadly and to comment freely upon varied aspects of labour–management relationships. Curiosity also appeared in the awesome and intrigued fashion in which they gratuitously commented on such peculiarities of the American way of life as consumerism, and egalitarian personal relationships. Linking the practical and personal in their reports was the ubiquitous Fordist philosophy that joint sacrifices for higher productivity raised everyone's living standards.

Specialisation and standardisation

Specialisation involved products, processes and occupations. The steels team emphasised the importance of specialisation in jobs. 'They were aware that by contrast with the tasks of the British craftsman, the work of the American operative is frequently repetitive, simple and apparently monotonous, and provides little opportunity for autonomous judgement' (Smuts 1953, p. 44). The teams, perhaps influenced by trade union members, were not over-impressed by this deskilling: 'Many of the teams ... criticised the failure to train a larger number of versatile craftsmen through apprenticeship programs' (Smuts 1953, p. 44). The valves industry team recommended more routes into the British craft category, rather than its decomposition along US lines (AACP 1951, p. 6).

Overall, production in the USA linked mechanisation, worker mobility and higher pay differentials. Although the gap between the highest and lowest rated workers was much greater in the USA than in Britain, opportunity to move from low to higher skilled jobs was greater. The lower degree of British mechanisation meant that the craftworker seemed to need assistance from many more unskilled workers – 10 per cent of total 'engineering' employees in 1937 (Yates 1937) – who had no chance of moving into skilled jobs (Smuts 1953, pp. 45–7).

This occupational specialisation was paralleled by specialisation in machinery. Production hardware seemed dedicated, in both senses of that word. According to one commentator:

> It was not the newness or the technical superiority of American machines which made the greatest impression, but the abundance of machines and power in relation to workers, and the widespread use of special purpose machines or fixtures, even for short production runs. (Smuts 1953, p. 40)

However, the AACP reports confirm the discrepancy between Fordist and Taylorist approaches specified in the previous chapter. A key aspect was the suggestion that Taylorist regulation of given work roles was more appropriate to most batch metalworking industry jobs where long runs were rare. The team report on the machine-tool industry noted that American manufacturers had gone a long way towards standardising their components; thereby removing much of the 'fitting' work in final assembly operations (AACP 1953, p. 48). In general, valve producers had managed to simplify product ranges, extend production runs, and correspondingly to specialise machinery because of the larger size of the North American market. So one large plant was able to 'specialise in variety' – making a wide range of valves in separate, large-volume departments. However, customer demands for variety impeded 'specialisation' and 'simplification' of product ranges in both countries. Standardisation meant, for example, one US valve manufacturer was making 128 different steel valves from a total of only 200 types of component. Yet the associated advantages of larger batch sizes was 'not as large as expected' (AACP 1951, p. 4).

As well as greater specialisation and lower average job skills US metalworking was also adopting Ford-style standardisation and simplification into batch production, even though the scope for these measures was restricted. AACP reports hint that British expectations of final fitting of components and sub-assemblies was less in evidence in the USA, because final-product tolerances – that is, dimensional precision – were less exact. Consequently, semi-skilled assembly of the final units was more likely in the USA. British observers also noticed that this made lower skills possible in the crucial machining occupations. Lower skills were also facilitated by the special-purpose tooling and fixtures that controlled the cutting tools. The extra investment in such equipment was made possible by the existence of the technical planning departments and the concentration on long runs of standard products. Standard products and fixtures obviated the need for the skills and experience otherwise necessary for frequent resetting of the machines. In some plants such standardisation through scale of production allowed one operator to work two machines simultaneously: a rare occurrence in Britain.

Organisation and methods

The extra bureaucracy of US approaches reinforced British scepticism. During the inter-war years many Taylorist and Fordist practices in the USA had been combined into centralised engineering planning and personnel departments (Jacoby 1985, pp. 99–104, 126–32, 195). In the steels sector '... the exercise of skill has been taken from the shopfloor into the pattern shop, the methods office and the brains of supervisors' (Smuts 1953, p. 44). Fordism's standard components for interchangeability of product assembly, and the Taylorist ideal of separate planning departments for tooling, equipment and machining methods, meant many specialist squads of technicians for draughting, planning and resourcing. The numbers of these, and of supervisory staff, contrasted noticeably with the generally lower proportions of administrative and technical staff in Britain (Smuts 1953, p. 44). The report on machine tools commented on the higher ratios of drawing and supervisory staff (British Productivity Council 1953, pp. 38–41); the valves team on the greater specialisation of planning staff and 'the advantage of sheer weight of numbers'. In one US firm 25 of the 115 employees were engineers. In another, the ratio of graduates, mainly engineers, was nearly 1:6. In valve-making firms such centralisation freed supervisors of much of the decision-making and administration borne by their British counterparts (AACP 1951, pp. 7–8, 9).

The British ambivalence to these arrangements echoed the turn-of-the-century attitudes noted in the previous chapter: 'Specialisation automatically produces complex problems of integration and planning and a sharp expansion of the need for supervisory, repair, maintenance, set-up, layout and design personnel' (Smuts 1953, p. 44). Centralised technical preparation meant careful engineering planning of jobs and detailed instructions, so that 'less attention needs to be given to the inspection side without harmful results' with 'few cases of keeping inspection records' (AACP 1951, p. 40). However, the implications for overhead costs troubled the British observers: 'We had hoped to find in America ideas for reducing paper work and administrative costs generally, but theirs are as heavy as ours' (AACP 1951, p. 52).

Labour inputs

The batch production origins and early applications of Taylorism made incentive payment systems the central device for raising worker effort. In practice Taylor's 'differential' piece-rate system was difficult to operate unless the conditions of task performance, materials and equipment could be kept standard. Otherwise workers would be unable to reach the strict threshold above which earnings increased with output (Hunt 1952, pp. 87–8). However, Taylorism placed more emphasis than other turn-of-the-century incentive pay schemes on detailed study and timing of task performance prior to rate setting. Wherever possible Taylorists also sought a redesign of the tasks and their execution; albeit changes restricted to discrete occupations, rather than the systematic redistribution of tasks across an entire production process of Fordist mechanisation.

As we saw in the previous chapter, Ford and his imitators worked backwards from a reshaping of the market and configuraton of the product, via component simplification and interchangeability, to achieve narrower ranges of final assembly and machining tasks with automatic handling and flow organisation. Fordism also differed from Taylorism by a determined elimination of payment by results. New jobs were 'time studied', but this was to achieve the fastest possible execution compatible with both accomplishment of jobs at earlier and subsquent stages of production, and with the setting of a fixed wage extra for each shift. Exacting time studies, direct supervision and the pace of the line regulated individual worker output, not the Taylorist incentive of extra pay for extra units produced.

In the USA the AACP teams encountered both Taylorist payment by results schemes and Fordist 'measured day work'. Yet while both systems were invariably based on time study in the USA, this was not so in Britain (British Productivity Council 1953, p. 43). Moreover, both the valve and machine-tool teams found that payment by results (PBR) schemes were restricted. Some US machine-tool firms were switching away from PBR to a 'points system' involving evaluation of job requirements and individuals' capabilities. Amongst other advantages such arrangements gave:

- greater consistency in settling pay between different jobs;
- more detailed control of excesses;
- productivity improvement through intensive study of operations involved in time-study methods.

Payment by results was found to be even less prevalent in valve-making firms. This team admitted that incentive schemes were not appropriate to mass production industry, but they had expected 'to find them almost universal in the valve industry, esecially as incentive schemes are widespread in the UK'. However eight of the fifteen firms visited had no incentive scheme; and three firms had established standard times for the output of every job with a high day rate for attainment (AACP 1951, p. 12). Both the specialist team studying the North American practice of industrial engineering, and the diesel sector study detected a 'trend away from financial incentive schemes and towards higher basic wage rates' (British Productivity Council 1953, p. 39).

Yet the machine-tool team asserted that 'monetary incentives do result in greater output per man-hour' and that 'The use of monetary incentives in the machine tool industry should be encouraged' (British Productivity Council 1954, p. 33). It seems likely that the British were trying to reconcile their own deep-seated prejudices about the virtues of financial motivation with the patent evidence that US firms were replacing some Taylorist forms of such notions by Fordist controls. A similar, if less specifically emphatic, predilection for incentive pay schemes came from the valves team. Despite their own evidence of a shift away from such arrangements, their criticism of the comparatively narrower wage differentials in Britain was complemented by a declaration of the virtues of financial incentives as a general principle (AACP 1951, pp. 10, 13).

However, there was a prescient awareness that 'there is a danger in Britain that motion and time study may be considered primarily as means of setting up an incentive scheme' (British Productivity Council 1953, p. 38). The machine-tools team also identified, implicitly, the difference between non-Taylorist piece-work schemes and Scientific Management ones. Incentive schemes should 'be more than a means of operators earning more money, they should be integrated into a control scheme whereby all excess costs are quickly shown up to management, who must then decide on, and take the necessary corrective action' (British Productivity Council 1953, p. 33). The dominant impression of the British observers was that American management had largely achieved this approach (Smuts 1953, p. 51), and that US unions either accepted or acquiesced in the use of PBR based on work-study techniques (Smuts 1953; AACP 1951, p. 12). Most of the British visitors probably recrossed the Atlantic believing that the overall superiority of the American way of manufacturing had triumphed over recalcitrant workers and union hostility to scientific management of incentive pay schemes. However, independent case studies suggest a different reality on the shopfloor.

Shopfloor realities

Donald Roy's classic 1940s study of payment by results and effort bargaining was replicated in the 1970s by Michael Burawoy (Roy 1952; Burawoy 1979). Burawoy's shopfloor studies showed that, even then, machinists on PBR were still able to manipulate 'official' time standards to get levels of output to match their own pay and effort expectations. The difference between the two periods was that the time study scrutineers who roved the shop in Roy's day had been replaced by more sophisticated, office-based industrial engineers with wider terms of reference than narrow task speed-ups (Burawoy 1979, pp. 162–70). Already by 1948 employment in US manufacturing industrial engineering departments was twice that of 1945. Thereafter such growth was confined to smaller and medium-sized plants. By 1957 a majority of such personnel was engaged in tasks other than time study and work measurement; with 18 per cent of this majority being directly concerned with general production methods and layout (*Factory Management and Maintenance* 1957, p. 12). In the theory proposed here, Taylorism was being overtaken by Fordism.

However, 67 per cent of US direct workers in manufacturing in 1957 were estimated to be still on Tayloristic 'premium' rate schemes (*Factory Management and Maintenance* 1957, p. 24). The 1940s and 1950s should, then, constitute an acid-test for the success of Taylorist effort regulation. For this was the high-point of such schemes in the USA. The study of Roy and others, confirms that the theoretical efficiency of Tayloristic PBR was limited by the ability of workers and their workplace union representatives to turn such arrangements to their own advantage. Workers restricted output because they believed that managements would cut the rate of pay if they reached the upper output targets consistently. Foremen colluded with workers' restrictions, because they feared that individual instances of maximum output would lead to pressure by middle management for all supervisors to achieve such rates. Union shop stewards were able to show a public face of opposition to restrictions while maintaining them in practice. In one plant, in line with the attitudes recorded in some British productivity reports, the official union organisation, the United Electrical Workers, actually mounted a campaign against output restriction (Collins et al. 1946, p. 7). As Carew has shown, US unions were, at best, reluctant co-operators in such efficiency schemes (Carew 1987, pp. 140–1).

Some reports recognised that union policies, such as those of the International Association of Machinists, were seen as formal restrictions of time-studied machine output (Smuts 1953, p. 51). Or, in the more dramatic prose of the Vice-President of the Kearney and Trecker machine-tool firm: 'When a machine can turn out 19 parts an hour, and the shop steward tells the operator 11 per hour is enough, we have in that simple fact a threat to our whole standard of living' (Trecker 1947, p. 13). Less partisan management observers recognised that the problem with Taylorist PBR schemes was that official and tacit union impediments arose from a whole complex of methodological and administrative weaknesses. Time studies were over-complex or there were insufficient data and a lack of standard conditions. Studies also tended to depend on the observer's experience for their accuracy, to be based on the most experienced or faster operators, and to make insufficient allowance for fatigue, personal time and delays. After the studies managers would invent slight, retrospective, changes to methods to correct 'loose' standards; while, from the workers' point of view, complicated incentive plans and formulae made computation of earnings difficult, and the bonuses that did accrue were not always shared properly with employees (Yulke 1947, pp. 28–9).

Rather than correcting such dubious practices a simpler solution was to abandon PBR completely and adopt Fordist 'measured day work' (MDW). Manufacturing managers thought MDW superior to PBR because it reduced the number of written grievances from unions. Savings on the direct supervision needed for PBR schemes were offset by the need for more inspection and quality control staff to check products and stem scrap rates. However, the authority added to supervisors by MDW made them identify more with managers (Schotters 1947). By contrast, the workers' grievances generated by PBR schemes were a potential advantage to unions since it was 'demonstrating the benefits of the union to employees' (Yulke 1947, p. 36). So, just when British representatives saw US practice as centred on Tayloristic time study methods, the main trend in the USA was away from such PBR towards Fordist approaches.

Bringing it all back home

Lacking detailed contemporary studies of specific UK sectors it is difficult to assess how far the productivity missions and other related schemes achieved a transfer of American practices. Melman complains bitterly that his 1956 proposal for standardised machine-tool production was rebuffed by the British authorities (Melman 1983, pp. 12–13). Protagonists cited specific advances made by individual firms as a result of adoption of both more specialised facilities and standardisation of components and sub-assemblies in some sectors such as diesels (Hutton 1953, pp. 212–13). The AACP claimed that, in Britain, as early as 1952, departmental and plant specialisation, together with movement towards flow-line production, was underway in the manufacture of internal combustion engines. Amongst diesel producers simplification of designs, new handling equipment and product specialisation were also noted. Although it was admitted that: 'most of the [diesel] industry works by batch or single job mainly because, in contrast to the American industry, it is dependent on many varied and mainly small overseas markets' (AAPC 1952, p. 34). Such Fordist reforms were, anyway, more than matched by the spread of Tayloristic applications. British particpants, with the odd astute exception (cf. Fletcher in Carew 1987, p. 146) saw the adoption of incentive payment schemes in the latter sector as an 'advance' (AAPC 1952, pp. 33–4).

Incentive pay: the acceptable route to Americanisation

Fordism would have been facilitated by larger markets, but these were restricted by the political failure to establish a trans-European trade zone. Longer-term and detailed planning to rationalise plant and equipment conflicted with the urgency of required changes. It was soon clear, therefore, that only some of the American productivity techniques could be implemented in the short term. Others were consigned to an indefinite longer-term future. 'Scientific' incentive pay schemes were a handy, and seemingly legitimate, remedy for this vacuum. In 1953 Hutton – an AACP apostle – recommended a distinction between fundamental changes in industrial organisation, which required longer time-spans or more structural reforms, and reorganising existing plant and methods for short-term productivity gains.

Management guides echoed this advice. They emphasised that improvements through PBR techniques, based on time and work studies, could be implemented quickly and at relatively lower cost than major technological and triple 'S' changes (Currie 1954, p. 679; British Institute of Management 1956, pp. 1–23). Work study, a British proponent argued, will:

> reduce the amount of capital required . . . or even . . . postpone the necessity for such expenditure. Because short-term improvements are always desirable . . . it is particularly attractive. (British Institute of Management 1956, p. 2)

Thus Tayloristic effort regulation was sold on the basis of the same expedients that had partly discredited Taylor's own ideas forty years earlier in the USA: high and speedy returns for low investment costs – a trend which has been paraphrased as an 'emphasis on the cost-cutting side of Taylorism, rather than on its output maximization side' (Merkle 1980, p. 235). However, the industrial relations of PBR based on time and work study – particularly those relating to effort bargaining – differed significantly between Britain and the USA. In the latter country the institutionalisation of bargaining by the legislative measures of the New Deal and its aftermath had not removed the managerial prerogative to define and propose changes to work practices. The legalisation of bargaining kept most union influence of these matters covert. Pay rating schemes and their standards were frozen for the duration of one to three-year contracts – legally binding management and union. Workers' dissatisfactions and protests did not disappear but were channelled into illegal 'wildcat' strikes. Or, more typically, dissatisfactions were expressed in the retrospective filing of official union grievances alleging management breaches of contracts for individual workers (cf. Jones 1984).

The role of shopfloor unionism

In Britain the boundary between management and union jurisdiction of these matters remained negotiable and contestable from plant to plant. In metalworking, the most formal regulation was in national agreements; but these mainly covered general principles. Increasingly, after the Second World War, shop stewards took up disputes over PBR and effort regulation at the plant level. In the post-war period general opposition to time-studied PBR was absent. Indeed, at the national official level, in the Trades Union Congress, as in the USA, it was formally accepted. However, the main union in metalworking, the Amalgamated Engineering Union, remained ambivalent: its core membership was skilled workers on jobs that were difficult to convert to PBR (Jefferys 1945, pp. 210, 256).

Moreover, national acceptance was premised on the tacit belief that the implementation of PBR could be modified to workers' advantage by detailed bargaining at plant and shopfloor levels. The post-war 'compromise' between labour and capital, the beginnings of corporatist regulation, the related enhancement of the legitimacy of organised labour, coupled with conditions of nearly full employment, all gradually increased the confidence and capabilities of shopfloor bargainers to exploit PBR schemes to the advantage of workers. Shopfloor unionism was becoming a stronger feature of British industrial relations, influencing the conduct of work and time studies, effort standards and pay rates. The same forces also gave greater scope for discontent with existing or changing PBR to be expressed in local collective action.

Transplanted into a cultural milieu of greater union and workgroup control over working practices, in a period of growing accommodation to union influence, Tayloristic techniques took forms that were plainly inconsistent with the decision-making prerogatives of scientific management. A Birmingham University study of five Midlands companies adopting work study found that:

> All of the firms made some effort to consult trade unions... Some firms trained trade union representatives in work study but kept them at work on the shop floor... others even trained trade union representatives from the shop floor and transferred them to a work study department to look after worker interests. (Davison et al. 1958, pp. 55–6)

American managers were already voicing dissatisfaction with labour's exploitation of the inconsistencies and subjectivity of time study in the 1940s. Yet it was not until the 1960s that British managements complained explicitly about the same limitations. Well-publicised cases such as the move from PBR to day rates at the Glacier Metal Company (Brown 1962) prefigured national protests. These linked effort bargaining over PBR with plant-level conflicts; as in the government's influential Donovan Commission of 1968. The Engineering Employers Federation evidence to the Commission referred to 'workers who ... could increase their output, and hence their own wages, are restrained from doing so for fear of the consequences', while 'some employers have encountered severe restrictions associated with the introduction of new methods through work study' (UK government 1968, para. 66).

Without US-style legal restrictions, unofficial strikes were less controllable by employers and the official union organisation. Although it may have been chosen selectively, the employers' evidence on disputes during 1965 suggests direct connections between PBR and shopfloor conflicts. Between May and October of that year 58,375 work days were lost due to 135 'unconstitutional' strikes associated with other causes. Yet incentive scheme disputes totalled 128 with 131,055 work days lost; while 54 of these stoppages, and 63,477 lost work days, allegedly arose from disputed prices or times for PBR work (UK Government 1968, p. 470). The employers acknowledged that, because downward revisons of rates often did not accompany repeated minor task simplifications, the system of time and work-studied PBR, itself, might be faulty. However they baulked at more Fordist alternatives:

> It should not be assumed that the panacea... is... to be found in a universal movement from systems of payments by results to payment on a daywork basis. It is not the piecework system but abuses of the system which create many of the difficulties. (UK government 1968, para. 88)

Less committed observers saw time and work-studied PBR as incompatible with the social regulation and economic context of British industrial work.

Lupton, arguing for properly negotiated and measured day rates, showed that a virtuous circle of higher output and higher wages – based on PBR – 'holds only on the assumption that management is in control' (Lupton 1967). In reality – and similarly to the US experience – groups of British workers were fixing upper limits to output. They might also have power to bargain 'loose' piece-rates. Moreover, managers themselves might allow a slackening of piece or effort rates so as to retain labour. Full employment and powerful union bargaining vitiated both 'more use of modern techniques of work measurement' or 'tighter control of the effort bargain'.

Division of labour and cultural expectations

A more fundamental obstacle was the adjustment of PBR weekly earnings, not by reference to output or time allowed, but by adherence to customary differentials in the workplace (UK government 1968, pp. 618–21). Both the classical American plant sociology of the 1940s, and the Birmingham University study of Midlands firms, had identified custom as the crucial weakness of 'scientifically' based PBR. Sargent Florence and the Birmingham team claimed that inadequate job analysis was not necessarily the problem. It was more an incompatibility between 'scientific' measurement and the need for regulation according to some social standards.

Even after adding allowances for personal factors and fatigue, as well as special allowances for delays: 'the standard rate of work set by work study was never the same as, and always greater than, the rate of work that was established before work study began' (Davison et al. 1958, pp. 49–50). Even then, and assuming 'accurate work study and given rates of pay for the job', there were other difficulties. The broader problem was the socially engrained preservation of established relationships between earnings. Contrasting work-studied, with more traditional PBR showed that:

> an implicit but accepted upper limit on earnings in the older schemes ... [kept] relative earnings ... in the same proportions in any given factory ... Post-war changes have tended to remove the upper-limit on earnings in order to reduce the effect of restrictions on output ... [but] differences in earnings between groups have caused extreme changes in relative earnings and the wage structure ... has been upset. (Davison et al. 1958, p. 52)

The final paradox of Tayloristic effort control by PBR was that the schemes were least suitable for the skilled batch production work which seemed the most in need of rationalisation. An alternative would have to raise productivity in jobs producing longer runs partly by technical improvements to equipment and planning – rather than by incentive pay. But, as the Birmingham team and others noted, it was easier to apply consistent incentive schemes to these semi-skilled jobs. When applied to the semi-skilled the improved incomes created resentment amongst the skilled groups, at the relative cheapening of their labour: ironically, a narrowing of the differentials which the AACP missions wanted so much to widen (Davison et al. 1958, pp. 53–4).

Shopfloor realities – illustrations of implementation

British managements opted for PBR based on work study as preferable to wider-ranging schemes of rationalisation, but the persistence of unrationalised craft and workshop forms of organisation – and the corresponding shopfloor attitudes – reacted against partial Taylorisation. The heartland of British metalworking in terms of both industrial activity and shopfloor cultures was the English West Midlands (Terry and Edwards 1988). Two of the Birmingham cases illustrate the detail of these failures and the importance of workgroup values and loyalties rather than trade union organisation *per se*.

One firm combined work study – to break down the job into standard units for timing – with estimates by a craft supervisor of the numbers in each job. The implication in the latter arrangement, that direct observation by work study specialists was resisted by the workers, is supported by the researchers' observations. 'The first few months ... were marked by constant arguments between the work study staff and the craftsmen about the times allowed for jobs.' However, it is also likely that inadequate organisation was involved. The craftworkers on one operation complained that:

> their allowed times had been miscalculated by the first supervisor involved because he had underestimated the number of elements in the job. One section of ten craftsmen thought their supervisor was unable to spend as much time in the workshop as was necessary because he was overloaded with paperwork as a result of the change. (Davison et al. 1958, p. 123)

The contradiction between shopfloor cultures and the official 'productivity-mindedness' is illustrated in the conflict between workgroup egalitarianism and Tayloristic rationality. One aspect was 'a conflict between the craftsmen and the work study engineers about the merits of individual incentive bonus. The craftsmen decided, as a matter of trade union principle, that they could accept incentives only if a group bonus was paid.' Another factor was that a stabilising adjustment in the analysis penalised slow workers more than it compensated fast workers. 'The men did not believe that the stabiliser worked fairly ... though the faster workers were prepared to work to offset the slower pace of the others' (Davison et al. 1958, pp. 81–2). On both sections the craftworkers alleged that the voting system had been manipulated so as to get a result supporting the incentive scheme; and that '.... their union had not helped in the matter because the union wanted the incentive scheme to be installed'. Craftworkers 'felt unfairly treated when semi-skilled men earned more wages than themselves' (Davison et al. 1958, pp. 125–6).

The AACP missions had envisaged the Americanisation of British production as a streamlining of inefficiency through a combination of more specialised, productivity-conscious workers and better planned, more mechanised equipment. Even where the craft ethos was absent there was a gulf between the official vision and the ensuing reality of short-term PBR fixes, as is vividly illustrated by another of the Birmingham studies. In this plant metal products were cut to length on machines; staffed by operators with a semi-skilled classification; although they actually needed 'some skill in setting the machine and controlling the flow of materials'. The rationalisation plan increased their wages but eliminated the jobs of nine female assistants who had previously carried away the product to the next operation. Output rose while total hours and wages fell. However, scrap rates also rose significantly as a consequence of a general deterioration in the quality of work tasks:

> Before the change the men were there to correct the fault before too much scrap had been produced. After the change the men saw the machines only when they were picking up loads... There was also an increase in the work done... dealing with unserviceable output... [lifting into bins] regarded by the men as an unpaid extra output not allowed for in the bonus scheme... if machines gave much trouble, the bins were not cleared promptly enough with the result that the men had to climb into the bins to compress the scrap by jumping on it. (Davison et al. 1958, pp. 72–3)

In a judgement probably applicable to many British jobs that were partially rationalised on the basis of piecemeal Tayloristic techniques the academic researchers drily observed of the harassed bin jumpers: 'There was no doubt that the job had been changed from an easy one into a strenuous one and the men were not satisfied that they were getting a fair deal' (Davison et al. 1958, pp. 72–3).

The Italian way

At the end of the Second World War Italian industry was sunk within the remnants of the Fascist system of economic regulation. Its owning and managing classes were discredited by widespread collaboration with Mussolini's regime. Warfare and Nazi looting had wrecked much plant. By 1949 it was still rare to find productive capacity that reached pre-war levels (Jacobini 1949). A majority of the industrial working class, which had seen an absolute decline in its basic living standards (Turone 1984), was rapidly mobilised into the communist and socialist sections of the new union federations. In these circumstances, the Marshall Plan's aim for rationalised methods of work and industrial organisation – so as to turn metalworking into a producer of standardised military, capital and consumer goods with an extra 200,000 jobs (CISIM 1952, pp. 23–4, 46, 48) – lacked the degree of political and social consensus that had eased its application to Britain. Although their reasons differed, government, capital and labour in Italy were all ambivalent about the technical transformation of its industry along American lines.

The industrial relations trajectory

In the early post-war years, after the US-contrived collapse of the broad left and right-wing governments, De Gaspari's Christian Democrat-led coalition stuck to deflationary fiscal and monetary policies. These were preferred to the Keynesian expansion of demand, that the American authorities favoured for the mass consumption counterpart to expanded industrial productivity (Hogan 1987; Barkan 1986; D'Attore 1985). The Cold War politics of the Marshall Plan were predicated upon the isolation and exclusion of communist political and trade union influence. Capitalists and managers thus had their Fascist-era predilections for authoritarian labour management (CISIM 1952, p. 95) strengthened because US aid was conditional upon measures to exclude communist and socialist trade unions (Barkan 1986). To a certain extent the CISL and the UIL – 'free' trade union federations allied with the governing parties – played into the hands of this tendency. Although the UIL later broke with CISL in favour of a locally focused strategy (Carew 1987, p. 178), both organisations first sought involvement in national industrial and economic planning, rather than productivity enhancement in the workplace and productivity-related wage increases on which their US and UK counterparts focused. Ironically, therefore, they failed to play the union role presupposed in the US model.

The socialist and communist unions who remained in the umbrella union federation, the CGIL, could claim to represent the majority of the workers in industries such as metalworking. However, their effectiveness was blunted, not just by the unrestrained anti-communist campaigns of the employers, but also by the absence of institutional arrangements to promote collective bargaining. In the immediate post-war years the pre-war *commisione interne* were revived as channels for representing grievances to management. Unlike comparable voluntary arrangements in Britain, however, these institutions did not involve recognition of unions. Nor, as with the legal mechanisms of plant bargaining in the USA, were there effective independent, external regulatory checks on employers' abuses.

As the Cold War period progressed, success in collective bargaining continued to elude the CGIL. Factory militancy during the Resistance, and in the unionisation campaigns at the end of the war had been centrally directed. The union continued to prefer a centralised focus in its dealings with employers until 1955. Then defeat for its candidates on the large FIAT *commisiona interna* finally encouraged a switch of emphasis to the strengthening of local bargaining. In the interim, FIOM – the CGIL's metalworking section – was unable to win any new contract from the employers' organisation between 1949 and 1956 (Barkan 1986, p. 46). The success of the employers' intransigence – often coupled with victimisation of activists and *de facto* bribery for non-militancy – and the ineffectiveness of national union strategy, contributed to steep declines in membership. Between 1950 and 1959 the number of members of FIOM fell by 404,000 to 185,000 (Barkan 1986, p. 47).

The significance of this industrial relations climate for the ways in which Italy applied Marshall Plan rationalisation can be appreciated by the contrast with Britain. Italian unions retained some influence on PBR schemes, such as a provision for protected earnings because of external disruptions to performance; and in Britain both employers and national union officials could oppose detailed shopfloor bargaining until the late 1950s. However, even under Conservative governments there was a *de facto* accommodation to labour's interests. There was no overt political split amongst the unions, and productivity schemes – such as work-studied PBR – tolerated, and sometimes even encouraged, union involvement. In the different Italian environment, management and employers took advantage of their virtual freedom in decision-making, to implement the technical features of the Marshall Plan in a crude, unilateral, Taylorist fashion.

Business and production policies

Under the Fascist system of economic autarchy industrial efficiency had regressed. Mass consumer markets were ruled out by declining real incomes, and the state steered the bulk of investment capital to the armaments industry. Although there had been an active Taylorist school amongst the technocrats of the inter-war period, incentives for widespread scientific management and Fordist methods were lacking. National standards and institutions for scientific management of labour and production were absent. Practices were particularised and diluted in individual firms and distorted by the bureaucratic organisation they tended to generate (Jacobini 1949, p. 107). Production engineering as an integration of machinery operation and production was 'almost unknown in Italy' (CISIM 1952, p. 25) and there was a lack of interchangeability in basic components – such as bolts, nuts and screws – for some products. As in Britain, there appeared to be inadequate pre-specification of dimensional tolerances, consistent with a low level of Fordist production planning.

The AACP teams often stressed that, handling equipment and power-assisted tools excepted, the general level of machinery in Britain was comparable with US installations. In the late 1940s, for example, several British vehicle firms adopted the newer automatic transfer lines – actually pioneered by British engineers in the 1930s – only a little after their American counterparts (Lewchuck 1986, pp. 152–84). In Italy, however, by the time of the technical assistance phase of the Marshall Plan in the late 1940s, the essentials of Fordist production were still often scarce or non-existent. By US standards most sectors were technologically very backward (CISIM 1952, p. 24).

Equipment for tracing, stamping, assembly, welding, presswork and turning were imported *en bloc* from the USA by some of the larger firms through Marshall Plan agencies (D'Attore 1985, p. 74). The volume of capital reinvested in machinery has led to assessments of the engineering industry shifting decisively from being an assembling, to a Fordist continuous flow industry. The scale of the transformation was epitomised in the contemporary slogan of 'less maccaroni, more machinery' (D'Attore 1985, pp. 70, 75). However, such a shift could apply only to certain mass production sectors. More generally, it seems, the industrial eclecticism typical of the Fascist period recurred.

There was a continuing dualism between the advanced and recapitalised sectors and more backward industries. Yet the former, often mass production firms, were still obliged to sell abroad as they had under Fascism, because of the limited purchasing power of most wage earners. In 1951 *per capita* consumption had still only reached a level equivalent to 1929 (D'Attore 1985, p. 78). The employers themselves were the cause of this failure to complete the virtuous circle of Fordism. Even amongst some of the model firms participating in the schemes promoted by the parastatal National Productivity Council (CNP) of the Italian Marshall Plan, there was sweating of labour by 'off the record' payments for extended hours and intensive working of low-wage apprenticed workers. Some craft-skilled jobs were reorganised into semi-skilled machinist and assembly work.

The Marshall Plan years also saw the beginnings of systematic job evaluation in some Italian firms (Barisi 1980, p. 11). Yet, in other cases, 'model' firms cut costs simply by refusing to recognise turners and maintenance workers as skilled categories for pay purposes (D'Attore 1985, p. 70). Incentive payment schemes, which were much closer to traditional piece-work than scientific management systems, spread on a scale hitherto unknown in Italy, a phenomenon whose apt popular label translates as 'ragamuffin Taylorism' (Turone 1984, p. 178). Significantly the most important initiative within the CNP's *Training within Industry* programme was the Taylorist scheme to remove supervision from the most experienced of the workgroup and make it the responsibility of a new category of technically proficient management appointees (D'Attore 1985, p. 82).

Conclusion

In the late 1940s and early 1950s, on the eve of the industrial stabilisation that underpinned the long post-war boom, a well-defined 'production paradigm' – with significant national variations – was crystalising in metalworking plants in both the USA and the European countries discussed here. The assembly and transfer technologies and pre-planning of interchangeable parts that constituted the production side of Fordist standardisation were being integrated into leading mass consumption goods industries such as automobiles. In other sectors – where narrower markets allowed only batch, rather than mass production – elements of Fordism, such as standardised components and fixtures for machine-tools, were adopted, but in piecemeal fashion. Fordist innovations were limited by the small size of markets and managerial conservatism. Where Fordist techniques were impossible or disdained, various Taylorist methods were adapted to fill the worst efficiency gaps. The most prominent of these were incentive payment systems based on time and/or motion studies to raise and maintain worker output; rather than the set times, machine pacing and high day-rates bequeathed by Ford.

In the USA, by the early 1950s, an amalgam of Fordism and Taylorism was evolving into procedures of production control dictated by the industrial engineer. This descendant of Taylor's work planning specialist now sought to integrate the regulation of worker effort through time-studied PBR with production scheduling techniques derived from flow production systems. However, the classic shopfloor sociology studies and the evidence cited above show how these methods could be frustrated. In key jobs, such as the setters and operators of machine-tools, the mixture of Fordist product planning techniques and Taylorist output incentives was a highly unstable arrangement for procuring high, low-cost, output of varying products. Moreover, the British context was even more volatile than the American industrial relations environment of specialised, segmented jobs regulated by legally based management decision-making prerogatives.

The Marshall Plan ethos shifted many British managers' production paradigm more decisively towards greater component standardisation, pre-planning of work, more automatic aids to machining processes, and the virtues of long runs of simplified products. Yet many firms could not specialise in simplified products because they lacked large homogeneous markets. It was easier to adopt piecemeal Taylorism in modernised terminology.

> The single major managerial development . . . has been concentrated in the area of production and on the lowest level of the managerial hierarchy. It is old fashioned work-study, bearing the new name of 'Organisation and Methods' . . . it represents a managerial movement that was revolutionary in American industry during the first decade of this century – but had been routinised into standard operating procedure in the United States by the 1920s at the latest. (Granick 1962, pp. 253–4)

Many firms were unwilling, or slow, to move to the kind of direct control over work tasks involved in either a Fordist segmentation of jobs, or a rigorous Taylorist regulation of existing occupations by work and time-studied incentive payments. Even in the mass-product automobile sector measured day work, paced by automatic machinery, was only consolidated in the 1960s. More generally, labour expectations constituted a significant constraint on the scope of managerial control over the work process because of an interrelated complex of social relationships: the emerging liberal corporatist accommodation with official trade unionism; the inheritance of deeply rooted shopfloor cultural norms about piece-work; and the growing strength of plant-level unofficial unionism.

Italian metalworking was also partly re-equipped and rationalised with the help of Marshall Plan provisions in the late 1940s and 1950s. However, the managers responsible made their changes in yet another kind of political and industrial relations environment. As Carew has explained, the Marshall Plan impact was patchy and lacking even the political coherence of corporatist leadership between unions and the state which was emerging in Britain (Carew 1987, pp. 162–3, 176–9, 213–14). The unions gave priority to national-level bargaining. US influence and state policies backed the exclusion and purges of communist unionism. Authoritarian traditions persisted amongst managements. All these contextual factors gave managers much broader scope for control over the work process.

Key Italian firms adopted North American standards of production technology. Ironically, however, the very weakness of organised labour in the factories, encouraged managers to eschew the US-style trade-off, between task determination and high wages, in favour of cruder and more exploitative methods of raising output. Despite the enormous financial and propaganda weight of the Marshall Plan campaigns, in all three countries basic workshop social institutions persisted and coexisted, albeit uneasily, with the new technical infrastructures associated with the changing factory production paradigms. In the next phase of factory evolution, however, this unstable triangle of institutions, technology and managerial paradigms was shaken by a new force in the technological trajectory: the arrival of micro-electronic computing power.

II Technologies of control

Overview

As chapter 1 explained, the analyst of industrial change has to avoid lapsing into either technological or socio-economic determinism. Over-emphasising one of these aspects means overlooking the special contributions that the other makes to actual developments. Moreover, the technological and the socio-economic spheres may interact in highly specific ways to influence the operation of the factory. These interacting forces include: intrinsic changes in technologies, which may be amplified or subdued by the sellers, the users or government policies; the micro-economic calculations of participants and the macro-economic conditions of an industry or country; the socio-political institutions – such as the organisation of employer and labour interests and collective bargaining systems; plus institutional and cultural forces, crystallising in managers' production paradigms. Such forces have their own trajectories, or histories which interact in complex, variable and often unpredictable ways, with sectoral, regional or national variations.

The remaining chapters examine the trends towards computer integrated production systems, what has become known as Computer Integrated Manufacturing (CIM) by pundits and commentators. As part of the increasing micro-electronic automation that is becoming such a central feature in larger factories it is tempting to regard CIM as some necessary logic of technology; or at least as a combination of technological and economic rationality. The aim of the next two chapters, 4 and 5, is to supersede such deterministic views by showing that the early evolution and adoption of computerised production control took specific forms, for specific purposes; and because of some very special conditions. From the start, futurist technologists and management pundits saw the computerisation of individual machine-tools, Numerical Control (NC) technology, as the first step in an inevitable progression to the fully cybernated factory. Actual adoption and diffusion, however, has not been an inexorable process. One homogeneous technological concept has not materialised.

Rather, there have been several competing possibilities. Second World War science made computer technologies a plausible proposition for manufacturing processes in the 1940s. However, their dissemination and utilisation has involved variant and sometimes rival types of application under the competing, but overlapping conceptions of Taylorist and Fordist production paradigms. The evolution can be shown to be more chequered and competitive, more a contingent than a deterministic Darwinian evolution.[1] In this way the socio-political forces, and their different national trajectories, can be more fully clarified than by assuming inherent technological potency and overriding business rationality. But contingency does not mean randomness.

As later chapters will show, until the advent of CIM installations such as Flexible Manufacturing Systems (FMS), specific conditions have worked against complex, potentially Fordist, forms of computerisation in metalworking. These conditions have favoured instead an unstable – and possibly fragmenting – Taylorist combination of conventional human skills and individual computerised machine-tools. Why did this discrete-machines approach win out, and why has it rendered the Taylorist paradigm unstable? The answers lie in four influential conditions responsible for prevailing usage of computerised machine-tools. These are: (i) the nature and distribution of technical skills with which the technology must interact; (ii) the administrative framework of the plant within which the machines and people must interact; (iii) the politics of interest representation and realisation within the organisation; and (iv) the sets of national institutions underlying and shaping the conflicts of interests – which create national trajectories in the socio-technical organisation of the metalworking factory.

For ease of exposition the interaction of technological evolution with these conditions will be presented in two chapters. The first (chapter 4) is restricted to the details and significance of the technological changes culminating in NC, and the issues of skills and administrative organisation. The second account (chapter 5) describes the role of 'plant politics' and the national institutional contexts.

4 Technological evolution and the pathology of batch production

For theorists of a shift to post-Fordism, Western industrial stagnation since the early 1970s reflects a generalised 'crisis of Fordism' in mass manufacturing. However, from another angle, the 'crisis' is as plausibly viewed as one of manufacturing inefficiency in batch production, that is, sectors unable or unwilling to adopt the Fordist paradigm. This chapter and the next describe the nature and extent of the automation of batch production through the technology of numerical control (NC) and related forms of computerisation. Initially a controversial technology, NC became the spearhead for predetermining production operations through computerised data and control mechanisms. For these reduced the range of manual controls and decision-making in the operation of machine-tools. Some of this story is already well recorded. Particularly important are the detailed accounts, from contrasting viewpoints, of David Noble (Noble 1984) and Charles Sabel (Sabel 1982; Piore and Sabel 1984). The aims of this chapter are to analyse whether NC represents a continuation of Taylorism and the onset of the cybernated factory – Nobel's thesis – or, according to Sabel, a successful break with Taylorism, and the beginnings of a post-Fordist industrial system.

Writing from a labour control perspective, David Noble argues that NC was developed and diffused in North America as a result of a concerted attempt by the US military-industrial complex to deskill and control recalcitrant skilled labour. NC, he says, was promoted for use by large corporations to the detriment of more skill-dependent alternatives. Sabel, on the other hand, points to the uneven adoption of NC, its distinctive technical advantages, and eventual successful adaptation to skilled work in small and large firms, in its more versatile computer numerical control (CNC) derivative. NC therefore signifies these writers' broader assumptions about either continuity or discontinuity in recent industrial change. I shall argue below that while the core of Sabel's version of developments is convincing, both his, and Nobel's more deterministic account, leave several issues unresolved.

Their discussion of the genesis of NC is limited to assessment of the merits of an alternative, contemporary type of machine-tool control technology known as 'record playback' (Noble 1984, ch. 7; Sabel 1982, pp. 64–5). Neither Nobel nor Sabel considers NC in relation to the range of other possible solutions to the problems of adapting batch production to factory principles. This chapter tries to do this by analysing NC adoption as a process of evolutionary competition with other alternative solutions. The focus of both Nobel and Piore and Sabel is on the technology's effect on the tasks and skills of the worker. This is certainly central to manufacturing production, but such a focus also abstracts from the bigger framework of the factory as an economic, social and technical entity in its own right. The wider context in Sabel's explanation of divergent uses of NC is, essentially, the distinction between Fordism on the one hand and, on the other, small-batch production of varied products for fluctuating markets. Fordism has a very general definition as the opposite of small-batch production; that is, as manufacture of long runs of undifferentiated items for stable markets.[1]

The present book also acknowledges that, as an overall factory structure, Fordism is most fully realised in the manufacture of standard products. However, in addition, this definition includes flow production methods and the pre-specification of product dimensions. Fordism is a paradigm which strives to organise the factory as a whole. It is more than a simple distinction between batch and mass production levels of output. Sabel distinguishes between the concentration of skills for large-batch production and decentralised skills in small-batch operations. However, except perhaps for the less typical, large-volume metalworking businesses, this distinction is unlikely to be very helpful in assessing the impact of NC.

For the latter has almost always been seen as a solution to efficiency problems in small-batch production – which constitutes the bulk of metalworking output.

Did NC succeed because it realised Taylorist principles, as in the labour control view, or because it dynamised the craft alternative to Fordism, as Sabel's perspective suggests? If one of these logics, rather than the other, guided the implementation of NC then the present patterns of factory evolution will differ accordingly. However, it has become increasingly apparent that broader institutions of the socio-political sphere shape core changes in the micro-economic, technological and organisational structure of industry. Viewed as 'societal effects' these institutions and activities have been shown to shape the organisation of work and production into distinctive national models (Maurice et al. 1986; Sorge et al. 1983). Earlier chapters have already shown, how such societal effects must also be considered as major influences on the implementation of production paradigms, and factory organisation more generally. Unlike Nobel, Sabel's account of NC adoption does include these influences. However, he cites their impact mainly as contradictions of any unilinear and deterministic logic of deskilling and labour control.

The nature of the product market is, for Sabel, the direct cause of whether NC work will be craft-based or hierarchically controlled. Masses of standard goods are taken to mean stable demand facilitating centralisation and hierarchy of decision-making, while a mix of small batches is said to mean variable tasks and some craft autonomy. However, the ultimate determinant is characterised as: 'the balance of power between labour and management, as well as ... the relations between different work groups' (Sabel 1982, p. 70). But what exactly influences this balance? Is it simply the arbitrary outcome of individual workplace struggles; which would invert the labour control perspective's determinism in favour of contingency? If not, then the extent of the influence of institutional conditions and the differences in national structures of power and interests must be identified.

This, and the next chapter, elaborate on Sabel's insights and contrast the intrinsic dynamics and permutations of organisation and technology – a potentially universal force – with the specific national patterns of institutions and interest group influence in Britain, Italy and the USA. The task of the present chapter is to identify the range of techno-organisational options for improving small-batch production, from which NC emerged in the 1970s. We shall find evidence that NC, and its offspring CNC, won out over competing alternatives partly because these alternatives required a more radical Fordist approach than the quasi-

Taylorism harboured by most small-batch managers. This is partially consistent with Nobel's claims. On the other hand, partly confirming Sabel's thesis, CNC matched not only small-batch methods of production, but also the skill resources already deployed in such firms. To resolve the competing claims answers are needed to two questions. How far has the eventual success of the NC option derived from such intrinsic technical, or micro-economic compatibilities? Secondly, does this still leave a role for the social institutions in which skills and managerial characteristics are embedded?

The limits to conventional mechanisation

Fordist methods – dedicated products, processes and sub-divided task skills – spread unevenly into Europe through the Americanisation programmes of the 1940s and 1950s. These methods were not universally applicable. They worked up to a point for firms and industries that could rely upon controllable demand and stable markets. But most metal components were not mass-produced in thousands, but in hundreds, tens or even single items.[2] The business economics of batch sizes hinges on two interrelated problems. Firstly, there is the technological potential for automating the machinery. Secondly, there is the organisation of the flow of work through different types of machine-tool. Engineering approaches to these problems are reflected in the evolution of machine technology since the nineteenth century.

At the centre of small-batch metalworking production is the machine-tool. As chapter 2 showed, this is both a creation and a creator of modern industry. The power-driven factories of the Industrial Revolution could not have been set up without Wilkinson's cylinder boring machine having been available to make Watt's steam-powered engines (Smiles 1863, pp. 178-82; Gilbert in Singer et al. 1958; Williamson 1968, p. 870). Evolving from the woodworking machines of the pre-industrial era, metal machining became pivotal for converting metal castings into components for the million-and-one kinds of machine, fixture, tool, fastening and hand instruments that are essential to advanced industrial economies. Only where components can be directly cast from hot metal, or stamped out by presses and similar machines, is cutting by machine-tools unnecessary. Simplification of work roles has been easier in the manufacturing stages that precede and succeed machining operations – metal forming: casting, forging etc., and often in assembly – than in the setting and operating of the metal-cutting machine-tools. Why have these operations persisted until the computerisation phase of the 1970s?

Most early machine-tools cut the metal piece to the desired shape by rotating or 'turning' it. So it is conventional to distinguish these lathes from the more complex types such as milling, boring and drilling machines. The latter remove metal by the horizontal, vertical or lateral application of a cutting tool rotating at high speed. Most of the mechanical characteristics of machine-tools were established in the nineteenth century, by innovations that began with Maudsley's slide rest and the gradual differentiation of milling and grinding processes from turning work (Jefferys 1945, pp. 13–15, 55–8; Rowe 1928, pp. 90–3). In the final decades of the nineteenth century greater product specialisation by firms on both sides of the Atlantic encouraged the use of chucks and jigs which positioned and controlled the workpiece during the cutting process.

In this way a considerable part of the machine operator's manual dexterity could be eliminated; provided, however, that product specialisation and batch size were sufficiently advanced to justify employing a department of skilled workers to make up the jigs and fixtures (Rowe 1928, pp. 90–2). Most of these innovations within the conventional mechanical technology were accomplished by the 1920s (Jefferys 1945, p. 20; Rowe 1928). The subsequent diffusion of electric power and high-speed cutting tools could only improve the speed of specific operations; not their preparation or control aspects. The productivity of metalworking batch production had reached a plateau by the 1960s. The variety of types of cutting operation in small- and medium-batch production needed variable tasks at the machine itself. Rationalisation could only come from two different types of change: either a technological substitute for the mix of human controls and interventions in individual machining processes; or a simplification and reorganisation of the types of work supplied to the machines. In practical terms a hierarchical Taylorist change or a lateral Fordist reorganisation.

A wider view

D. T. N. Williamson was an innovating British engineer, who participated in developments on both new Taylorist and Fordist technologies. By the 1960s he, and other perceptive protagonists, had recognised that the root problem of small-batch pathology was the incompleteness of factory principles in metalworking. Modern plants had not altered the basic structure of the nineteenth-century workshop. Small workshops produce varied batches relatively efficiently because the skills and commitment of technicians and machinists are embedded in informal communication networks. These provide a human feedback system to harmonise designs and methods, and to detect, correct and forestall errors. This workshop structure could have been effective if the twentieth-century expansion of batch production had been realised through increasing the number of such small production units. Indeed, this issue of size and organisational character are key themes to which we shall return in the discussion of Italian manufacturing developments in chapter 9. The dominant trend, however, has been for an expansion of the scale of workshop operations into establishments of factory dimensions; but without a corresponding rationalisation or integration of machining operations.

The main tendency was machine specialisation in the Taylor style. The workshop form of organisation was retained but work was sub-divided according to the type of machining operation required: a specialisation of functions and often, but not always, of tasks. Larger and larger numbers of small batches were sent through factories with machines specialised by function. Thus all the turning operations are done in one section or shop; all milling operations through lines of milling machines, and so on. Further efficiency was left to unreliable Taylorite attempts at skill specialisation amongst machinists, and the related regulation of the execution of their tasks. This functional organisation had two general effects. Firstly, operators, or setter-operators, became skilled specialists in a closely related range of tasks. They were indifferent to the overall product because they dealt with a wide range of components and operating variables, but only a specific type of operation – such as turning and milling. Each contributed only to a small fraction of the final product. Secondly, enormous delays and stocks of work-in-progress ensued. Components 'queued' for their turn on different machining sections. Consequently the actual metal removal phase was estimated to make up as little as 1–1.5 per cent of the total time a component was in production (Williamson 1968; McKeown 1981).[3]

By the late 1960s – in the USA at least – batch production operations were being pressured for faster delivery times by the end-users of the components and sub-assemblies that they machined. Yet, contrary to Fordist logic, speed-up through standardisation was limited by simultaneous increases in the numbers of product lines (Rader 1969, pp. 98–9). An important influence was the changing pattern of demand for military equipment. Military production in particular was a major source of manufacturing orders in Britain and the USA.[4] But weaponry requirements moved away from the mass production of aeroplanes towards smaller numbers of a variety of jet planes, missiles and other aerial armaments (Jones 1985a). Batch sizes of components were, therefore, reduced.

Competing solutions and persistent paradigms

As these various problems became critical in the 1960s, three different engineering solutions had been developed with plausible claims to advance factory efficiency in small-batch metalworking. One was NC machine-tools, which offered automation of much of the operator's decision-making and manipulative tasks. Integrated cells and group technology, the second and third solutions, shared a number of similarities deriving from the Fordist logic. The aim behind *integrated cells* was to invert the conventional allocation of parts to functionally specialised machining sections. Instead, different machine-tools would be combined together in cells, according to more restricted but complementary operating tasks, and earmarked for particular products or sub-assemblies of components. The cells could then be supplied and linked by Fordist techniques.

Group technology also aimed at rationalising the allocation of machining but sought this by detailed classification, planning and allocation of the components to be machined according to criteria of shape and size. As with integrated cells the conventional layout of machine-tools, according to functions, was to be replaced by grouping them to cater for particular classifications of principal components. This arrangement was to minimise resetting of machines and eliminate queuing. On its own terms group technology, or GT, was a radical – its advocates termed it a 'revolutionary' – prescription for batch manufacture. Both GT and integrated cells were certainly more radical than the mishmash of Fordist and Taylorist techniques promoted during the Marshall Plan period. Assessing how GT and integrated cells were eclipsed by NC helps, therefore, to clarify the nature of the technological evolution under consideration here.

GT proposed that machining sections would be composites of all of the types of machine-tool – milling, boring, grinding etc. – necessary to perform the operations on a 'family' of related components. Like Fordism then, GT sought to work backwards from the nature of the product to the most efficient organisation of the production equipment. Unlike Fordism, GT did not stray out of the given character of the components by attempting to simplify or standardise the product. GT had two important implications for work and the management of production. Firstly, the tasks of the machinists would be simultaneously simplified and widened. So, instead of being proficient in a wide range of, say, milling techniques – and the corresponding setting tasks – the operators need only be conversant with the milling techniques on a certain size or shape of component. On the other hand, they would need skills for a similarly narrow range on each of the tasks, drilling, grinding, boring and so on, required for the particular family of components. Because the range of workpieces would be more homogeneous, setting of the machines would be simplified. Key organisational aspects of the contemporary fashion for 'just-in-time' manufacturing can be traced back to GT.[5]

Secondly, however, GT presupposed considerable amounts of planning and administrative work. All parts to be machined must be graded and catalogued with individual part numbers. Design procedures were to be overhauled so that new orders could be met from the existing catalogue, or to minimise design work by allowing minor modification to existing parts. Finally, the greater interdependence of work tasks and machines within the groups meant that more detailed scheduling and timing of jobs would be needed. Inevitably, computerised data storage would be desirable for the huge stocks of information on which the system was based (Leonard and Rathmill 1977).

The essence of GT was variety reduction. The integrated cell approach, on the other hand, involved a less complex rationalisation. Nor did it aim to simulate mass production by grouping the components solely to achieve minimal resetting of the machines. Cellular manufacturing resembled GT in aiming to simplify the machining process by dedicating a group of machines to a similar and limited range of parts. The items to be produced, however, were loosely grouped according to the functional characteristics; e.g. gears, ratchets and sprockets as 'transmission items'. This kind of grouping was seen as conducive to making machining requirements similar and avoiding the time-consuming and complex classification procedures of GT's geometrical numbering schemes. What was distinctive, it was claimed, was the arrangement of the machinery as a cell of complementary machines. The composition of the cell amounted to decomposing the various tasks of the single all-purpose machine-tool to a variety of machines. Self-contained, 'stand-alone', machines were to be abandoned, and replaced by more specialised, yet complementary, tools. These were to be distributed around the cell, and utilised according to the precise requirements of the product range.

Despite an articulate lobby of expert technologists, few cases of integrated cells were installed in British factories. Yet the most advanced, the SYSTEM 24 at the Molins Machine Company, is of special significance to the recent evolution of the batch production factory. With its centralised computer control of the parts and tooling, SYSTEM 24 may be regarded as a direct ancestor of the high-tech – and potentially Fordist – Flexible Manufacturing Sytems (FMS), which are at the centre of the most recent developments in factory technology.[6] It was also the brainchild of D. T. N.

Williamson, then Director of Research at Molins. As an early developer, and proponent of NC machines at the Ferranti company, Williamson became a leading advocate of integrated cells during his time at Molins. Although the SYSTEM 24's computer instructed the dispatch of parts and tooling to the machines in the cell, workpiece setting was organised along more conventional Fordist job-simplification lines. Williamson himself went so far as to describe the system as a 'flexible transfer line'. Wherever possible work skills were simplified. Specialised female operatives – 'girls' in Williamson's revealing terminology – were sent bins containing instructions, the raw metal billets and machine fixtures. These were assembled on special purpose pallets and fed via a conveyor belt into storage racks, to await transfer to the machine-tools. Consistent with the practice of Fordist assembly lines the job times of the 'girls', and completion rates, were monitored and supervised from a central computer room, from where the mainframe computer also commanded the machining operations.

A most significant contrast between NC, on the one hand, and both GT and integrated cells on the other, is the consistency of NC with Taylorist practices and the latter two approaches with Fordism. In arguing for the standardising approach embodied in integrated cells, for example, Williamson derided the efficacy of excessive administrative controls and top-down instructions that we have identified with conventional Taylorist organisation of small-batch production (Williamson 1968, pp. 874-6). Although the philosophy of GT and integrated cells was compatible with the use of NC equipment, the paradigm supporting NC, as such, actually differed sharply. NC was seen as improving the speed of setting and operating of individual machine-tools, as well as enlarging their metal-cutting capabilities. Technological visionaries looked ahead to controlling entire sections, even factories, of machine-tools by computers: technologies which have subsequently been labelled as Direct Numerical Control (DNC) and Computer Integrated Manufacturing (CIM).

However, NC became publicised and initially adopted as a classic, self-contained, Taylorist device. It seemed to offer hierarchic control and simplified work tasks for individual work roles by buying off-the-shelf chunks of hardware and software. This situation of competition between rival systems has not been completely resolved. Yet, in many respects, NC has won out over the alternative solutions to small-batch complexity. This is despite the fact that its most successful applications undermine Tayloristic principles of hierarchy and control. Moreover, aspects of GT and integrated cells have reappeared in revised versions of the Fordist paradigm associated with Flexible Manufacturing Systems. But all of this takes us ahead of our story, which must first explain the chequered evolution of the NC technology.

NC: natural selection or social adaptation?

Technologically centred theories of the development of factory production resemble crude Darwinian conceptions of evolution. They assume that from a preceding technology new artefacts are developed to meet requirements that the previous systems could not handle efficiently. After a brief period of 'resistance to change' by vested interests, or irrational prejudices, the new technology displaces the previous one; because it is inherently better suited to the technical and economic demands of production.[7] The polar opposite types of interpretation (Braverman 1975; Noble 1984; Marglin 1976) replace technological with a socio-political determinism. In this view one specific technology appears as a replacement for another, largely, or exclusively, because it is more effective in realising a dominant politico-economic interest. Succession is also clinched by the dominant interest possessing the social and political power, and means, to force through the preferred technology; even in the face of obvious technical and economic shortcomings. Evolutionary terms and metaphors are sometimes deployed loosely in accounts of the technological diffusion (cf. Williams 1983; Rosenberg 1976, p. 174; Nelson 1987). However, a much more complex process can be revealed, if we adopt a more systematic evolutionary parallel for NC than either technological or socio-political determinism allows.

Biological evolution consists of an interplay between the dynamics of the species and the environment. The successful development and predominance of a given species might be ascribed to the overwhelming pressure of environmental demands. By analogy, the environmental demands on technology would be the politico-economic thrust of the dominant business interests. Alternatively, the emphasis might be on unique combinations of unfolding properties inhering and germinating in the species – technical advantages designed into the technology (cf. also Nelson 1987, pp. 13–14). However, post-Darwinian palaeontology would not ascribe the success of a species to just one of these factors. Indeed there is a very strong, contemporary, palaeontological perspective which argues that the predominance of one species over other, potential rivals involves immense complexity, and even contingency, in the interaction of biological specifics and environmental forces (see Gould 1991; and note 1 on p. 266). It would be implausible to suggest that mammals superseded dinosaurs just because their anatomy gave better means of consuming nutrients. Nor would a view of pure environmental determination – whether permanent winter from inter-terrestrial collision, or deterioration of sources of nutrition, and so on – be credible. The ascendancy of species arises from complex interactions between environmental and physiological factors. If the analogy is appropriate then technological succession can be ascribed neither to the immanent properties of a new machine or system, nor directly to the dictates of a dominant group's preferred environment.

For the evolutionary analogy to hold, a minimum requirement would be some mutual reinforcement between the socio-economic environment and the technological properties. An evolutionary treatment of the successful spread of NC machinery raises two particular issues. Firstly, as in palaeontology and other cases of technological succession, several solutions have been competing with NC to be successors to the massed sections of conventional machine-tools.[8] Secondly, there is the heterogeneous 'environment' of economic interests and concrete engineering practices; one not reducible to a single controlling interest, such as capitalist accumulation, military-industrial power, or Taylorist control imperatives.

Incomplete evolution

Mechanisation of the control dimension of machine-tools began in earnest when the rudimentary cybernetics, developed for the control of Second World War ballistics, were applied to tool and workpiece movement. By the late 1940s and early 1950s technologists had developed NC machines run by a separate controller – large mainframe computers. These took their instructions from numerically coded paper tapes, and activated servo-mechanisms on the machine-tools themselves by electronic signals. In this way it was possible for an engineering technician – or other numerate, computer-literate office worker – to write 'part-programs'.

These were long chains of instructions, that would eventually control the machine through 'closed loop' feedback principles; continually correcting the speeds and distances of tools and workpieces. NC had two clear differences from previous methods of automating machine-tools. One was the capability to change the control procedures – simply by revising or modifying the program – without much mechanical change. This facility did not need the jigs, fixtures and special-purpose machine-tools previously used for predetermined control of tool cutters and workpieces. The other difference was precisely its compatibility with general-purpose machine-tools. Automation of batch machining, not just its mass production offspring, was now achievable. The early historical development of NC has been well documented elsewhere (Mansfield 1971, pp. 187–8; Noble 1984). What has become increasingly doubtful in the last decade is whether its spread constitutes a technological embodiment of Tayloristic principles, as was first claimed. Many managers and technologists, especially in the USA, suggested that only 'machine minders' were necessary, and that morons, or perhaps even trained apes, could be taught to stop, start and monitor NC machines because all of their metal-cutting motions were guided by pre-planned computer programs (Fadem 1976, p. 10). These claims supported the interpretation of radical critics (Braverman 1975; Noble 1984; Shaiken 1984) of NC usage – as a Taylorist separation of execution from conception (Braverman 1975, pp. 199–205). However, the evidence of both early NC usage, and its subsequent chequered and heterogeneous usage, contradicts such a unilinear interpretation of progressive Taylorisation.

What Noble's detailed history of the early NC developments does prove is that in the first decade of its promotion the technology was technically complex, expensive and ill-adapted to the needs of the majority of relevant metalworking firms. The initiative behind the invention came from the US Air Force who were seeking a way of cutting the complex wing shapes of the new jet aeroplanes more quickly and reliably than with either the jigs and fixtures, or lengthy manual operation of conventional machine-tools (Noble 1984, pp. 101–6). However, there is disagreement over the most important aim arising from this linkage between military requirements and NC development. Noble, echoing Braverman, stresses its capacity for eliminating skilled, costly and troublesome labour. Sabel, on the other hand, contrasts NC's superior automatic manipulations of tools and workpieces, with the sheer inability of manually controlled conventional machines to cut the exotic, hard metals of the new jet plane airframes (Sabel 1982, p. 64). As the case of FMS will confirm in chapter 6, it is almost impossible – short, perhaps, of psychoanalytic hypnosis – for the outside observer to discover with certainty whether the rank ordering of the publicised advantages corresponds with the actual ranking made by buyers and users in their choices.

More clear-cut is Noble's account of the promotion phase of NC. Enormous sums were spent developing a programming language capable of expressing all the technical information required by the engineers and technicians in the workplace, that could be conveyed to the machine-tools. It is clear that the diffusion of NC in the USA was checked for several years, between 1957 and the early 1970s, by the failure of makers and users to agree on a standard and workable programming language (Noble 1984, pp. 142–3, 225–7). Diffusion outside of the aerospace sector was correspondingly slow, and attitudes were sceptical. Programming tasks required some aptitude or experience with 'high-level' programming languages, as well as access to expensive mainframe computing facilities. Reactions in Britain were also mixed.

Further adaptation

D. T. N. Williamson, while at Ferranti in Scotland, recommended NC in 1955 as 'the complete answer to the shortage of skilled machinists' (Williamson 1955, p. 152). Yet by 1968, when NC use was still largely confined to the aerospace sector, Williamson was discounting NC in favour of the integrated cell approach. He argued that resetting of the workpieces restricted the actual cutting times of NC so that, despite possessing some advantages over conventional machines, 'a really substantial reduction in cost is not one of them' (Williamson 1968, pp. 880–1). Independent British sources also saw successful use of NC as dependent on the organisation of machines into integrated or GT cells (Swords-Isherwood and Senker 1978). As late as 1979, on a visit to a specialist NC-using sub-contractor to the British aerospace industry, I was asked nervously by the general manager whether managers in other firms really did regard NC as worthwhile.

Yet around that time investment in single NC-type machine-tools expanded significantly. As well as the more complex contouring and five-axis machines originally demanded by aerospace firms, manufacturers began providing two-, and three-axis equipment, plus straight-line and point-to-point cutters (Gebhardt and Hatzold 1974, p. 53). Perhaps most importantly, the micro-processor advances of the 1970s superseded the original need for laborious programming in relatively high-level programming languages using mainframe computers. For these firms often recruited technical and computer staff for program writing. But their inexperience in machining techniques led to errors and time-consuming rectification during trial runs – 'prove-outs' in NC jargon – or, even worse, during actual production. Micro-processor electronics reduced the size and cost of computing devices, simplified programming functions in CNC machine-tools and eliminated a need for mainframe computing access. Part-programs could now be made at the machine-tool itself – although this by no means meant programming by shopfloor workers. CNC did, however, make it easier to draw on production workers' know-how: maintaining a Taylorist organisation by promoting them to part-programmers, or by delegating some programming tasks to the operators themselves.

NC sales had grown somewhat in the 1960s, and then slowed again. In most of the advanced capitalist economies the proportion of NC amongst all machine-tools remained below 1 per cent. Between 1975 and 1981 this doubled or trebled (Freeman 1985, p. 30). Most of the growth has been attributed to the switch by several Japanese machine-tool firms to production of cheap CNC lathes and more versatile multi-function machining centres, which were also equipped with the new computer controls. Western machine-tool manufacturers were caught unawares as Japanese supply of such machines increased ten-fold between 1970 and 1980 (Piore and Sabel 1984, p. 218). The Japanese industry became the world's leading NC manufacturer (Freeman 1985, p. 27) and the dominant force in small NC equipment in the USA. This expansion derived from marketing the machines amongst the smaller 'jobbing shop' metalworking establishments; rather than the big, sophisticated, computer-rich plants in aerospace and transportation equipment (Piore and Sabel 1984, pp. 217–18).

In Britain a similar shift was observable. A later investigation by Swords-Isherwood and Senker, of the GT-oriented firms they had visited in 1970, found that by 1978 NC had become the more productive investment. This was the case even where planning, supply and costings were geared to single machines; rather than the allegedly more efficient GT overall-production approach. The finding held even for firms that had previously introduced GT and 'some of those who had previously introduced GT seem now to be returning to a more conventional machine shop layout' (Swords-Isherwood and Senker 1978, p. 45). GT was also shown to have a distinct advantage over functional layouts of single machines only where there was little variability between parts and low complexity in their manufacture. In other words a combination of large quantities, simple parts, and minimum set-up requirements was necessary (Leonard and Rathmill 1977). The analysis had come full circle. GT was only effectively applicable where batch production resembled key features of Fordist manufacturing. Yet it was precisely to the problems of small-batch production that GT had originally been proposed as the solution!

NC evolved as an acceptable improvement in the factory organisation of batch production for both technical-economic reasons and for underlying social reasons. It could show more concrete immediate cost-savings than GT, or integrated cells, and it did not presuppose Fordist-style product and process rationalisations of the production process as a whole. NC's original hardware and software configuration was often inappropriate for non-aerospace small-batch manufacturers. However, its adaptation to the more general small-batch environment, largely as a result of CNC and Japanese modifications, made it congenial to the run-of-the-mill machine shop manager. In its early days, at least, NC was also consistent with the labour control element in the Taylorist paradigm. But the expansion phase of NC coincided with the more 'operator-friendly' development of CNC systems and allowed smaller firms to use them in Tayloristic or informal forms of organisation: an issue in its own right, to which we shall return in this and the following chapters. The thesis that NC became the domimant new technology only because it furthered the immanent logic of Taylorism in factory organisation is, therefore, difficult to maintain. On the other hand its success must be examined in terms of how it was used. NC/CNC was rarely accompanied by large-scale reorganisations of skills and work roles. It is not the superior design or technical potential of computerised machine-tools which clinched their acceptability but, seemingly, their compatibility with varying organisational forms. For example, larger firms could still be found using the newer CNC equipment with the original centralised programming arrangements of NC (Jones 1983).

In the last analysis practical success lay in the manner that NC/CNC was actually used. By a process of deductive elimination of the technical and economic factors that could have explained its success, we are left with compatibility with pre-existing socio-organisational arrangements, as the key variable. This factor is not equivalent to the much broader socio-political forces, notably promotion by the US military complex, identified by Noble. Such constellations of socio-economic power were important in the initial developments. However, as far as widespread, and successful, usage is concerned, it is the relative importance of technical-economic factors that needs disaggregating from the immediate social framework of the industry: administrative organisation and social institutions of interest groups. The rest of this chapter will establish the boundaries between these two causal influences in the intermediate area of skill utilisation. The next chapter will examine the relative influences of administrative organisation and the character of the social forces in the three national contexts of Italy, the USA and Britain.

Skills: the bridge between the technical and the social

Workers' skills are the medium by which self-regulating but contextually insensitive machine technology is adapted to the contingencies and priorities of the factory floor. Skills are a bridge between the social ordering of human productive forces and the inanimate means of production. For only a social context makes this or that level of quality, this or that use of raw material, economically or socially acceptable or significant; and only social backgrounds can constitute effective human skills (cf. Collins 1990). Because of this transverse role, both the technical and social characteristics of skills have to be considered in relation to NC technology.

Technical options and constraints

It was not excluding the involvement of workers and their skills which made NC, or more accurately CNC, the distinctive late twentieth-century development in small-batch metalworking production. This was one of the early claims made by NC proponents and has continued to be salient for some plant managers. However, the principal cause of the later success was that it became apparent that NC-automated processes could best improve technical quality of production and productivity when combined with certain human skills. What then were the precise advantages that NC usage provided and with what specific kinds of skill did they successfully combine? The three general and overriding advantages associated with NC/CNC machine tools are:

- mechanical flexibility – the workpiece and tooling equipment can be manipulated in ways that are beyond the scope of most manually controlled operations;
- repeatability – once successfully used, the software for a particular operation, the 'part-program', can be reapplied in exactly the same way;
- speed – the larger tooling stock and automatic sequences of cutting operations minimise the time taken to perform the range of tasks involved in machining a given item.

From these three general qualities stem, in principle, a variety of more specific economic and technical gains in particular production plants. These gains include: faster 'lead-times' between production planning and actual production; more intensive use of the capital stock of machine-tools; and reduced errors and scrap once a part-program has been 'proved-out'.[9] These advantages appear such compelling reasons for investment that technical-economic grounds might seem to be sufficient reasons for NC's successful adoption. This is no doubt partly correct, but only as far as its direct appreciation by plant management is concerned.

It is true that the early NC phase was accompanied by a strong current of opinion, amongst both academic commentators and industrial participants, that the NC operators could be reduced to a semi-skilled level of competence. The argument was that manipulative ability – 'sensory-motor skills' in the jargon – for cutting tool movements had been eliminated. So also, it was argued, was the need for shopfloor workers to interpret symbolic and written plans and drawings (see Hazelhurst 1967, pp. 6–14). Contrarily, for many firms, the high costs and uncertain reliability of the new machines led them to adopt a 'play-safe' policy: keeping workers with a craft level of conventional machining skills in NC operator jobs, as an insurance against costly but unforeseen problems. Where the nature of the products required considerable tool resetting then it made even more sense to use skilled workers for the setting and operating tasks. However, in other instances – especially where batch sizes tended to be consistently larger – setting and operating functions might, and may still, be separated. Thus one or a few craft-skilled workers could undertake the former role leaving semi-skilled workers as operators (Jones 1982; Whittaker 1990, p. 141).

The skills were still present on the shopfloor, but at one remove from the actual operation of the machine-tool. Nonetheless, even firms with this separation tactic for 'diluting' operator skills in the 1960s and 1970s often reversed this policy later. Sometimes this reversal stemmed from the countervailing pressure of trade unions (Jones 1982; Wilkinson 1982) – an issue discussed in the next chapter. But in other cases there is evidence that managers recognised that they had underestimated the need for someone with conventional machining skills to attend the NC machine (Swords-Isherwood and Senker 1978). The tasks requiring these skills were the monitoring of the cutting process and the fine tuning of the NC program.

Types of skill

Before examining the different social factors influencing skill application and NC usage it is also necessary to be clear about the nature of the skills involved. In the absence of a generally applicable classification of skill types[10] two main dimensions of skills can be identified in machining by NC and CNC technology. Firstly, there is the *adaptive-mechanical* dimension, in which considerable residues of craft mechanical skills are still needed to set, check and adjust tooling and the metal pieces and parts being machined. Secondly, there is the *representational* dimension. This is composed of artificial mental representations of the programming operation which correspond, in turn, to the machining process.

In the first dimension pre-set tooling, automatic rotation of tools and workpieces were already limiting the need for manual interventions in the 1970s. Often, however, the vital first trial of the program, the 'prove-out', needs adjustments before it is ready for normal production. Even during the normal operation phase, mechanical knowledge of appropriate tool movements and sequences of metal removal may be required in order to optimise the economy and quality of the cutting. Or, where a full and rigorous prove-out was not completed, monitoring by mechanically skilled workers would be applied to guard against catastrophic 'crashes' of tools and metals. Underlying these manipulative-sensory craft skills is conceptual knowledge: of geometric dimensions and methods of calculating the speeds of cutting tools, and the rates at which the metal component is 'fed' to the cutters.

In the second, *representation*, dimension of less tangible skills, such mathematical knowledge is repeatedly called upon in a tacit fashion. It overlaps with similar judgements exercised in a more conscious and overt fashion. It is true that a new program can largely be made up by transcribing data from manuals on metal-cutting speeds and feeds, or from records of previous worksheets and similar past programs. However, in all except the simplest cases the person writing the program will mentally translate specified information on dimensions, tooling and methods, and synthesise these into instructions in computer-readable form. These instructions will then be entered, as new data, into the appropriate medium – a paper or magnetic tape for NC controls, or the memory of a mini- or micro-computer in the case of CNC. All the time the program compiler should be 'envisaging' (see Grootings, Jones and Scott 1989; Cavestro 1989, p. 233) the actual course of the workpiece and the cutting tools.

In practice these two dimensions of skill, the adaptive-mechanical and the representational, are interlinked. Shopfloor workers, too, need to utilise the representational dimension: to 'envisage' the actual machining process and how it is symbolised in computer-readable forms. As a preliminary to program writing this form of symbolisation is more complete if it is based upon experience of the particular calculations and manual adjustments involved in setting tools and workpieces. Likewise in the adaptive-mechanical sphere, rectifying minor errors and optimising inadequacies in programs during the early machining runs, is easier if the person in charge of the machine-tool has understood the logic and sequence of the instructions encoded in the programs. Far from NC eliminating these skills it actually required their presence in more intensive, or less routine applications. Some of the practical problems of NC and CNC help to verify this proposition.

Compromising the Taylorist paradigm

The theory which Braverman and some of his followers took for reality, was that a fully prepared program would cover all the details and possible problems in the actual machining process. In practice both large and small difficulties were not – and still are not – captured in the software. By missing out a single character or command, programmers might unwittingly send the cutting tools to the wrong position during the metal-cutting operations. Programmers might forget or never have been told, that a particular machine ran at slightly different speeds to others of its type. They might program for cutting one type of metal when, in fact, another was being used. Sensitive monitoring might also be needed because of the state of the metal billets, from which the components are to be machined. These billets may also vary considerably – and unpredictably – in their hardness or porosity; and thus require faster or slower, or more careful cutting.

Skill needs vary with batch size. Large batches – long production runs – infrequent resetting of machines, and low levels of accuracy, quality of finish and materials costs, all may allow firms to employ operators with less formal training in recognised skills. However, where these factors tend in the reverse direction, managers are more likely to 'play safe' by keeping skilled workers to operate the machines. A wide range of studies have identified these or similar combinations of skill and technical exigencies (Batstone et al. 1987; Buchanan 1983; Burnes 1984; Cavestro 1989; Hazelhurst 1967; Jones 1982; Kelley 1984; Sorge et al. 1983; Swords-Isherwood and Senker 1978; Wilkinson 1982; Wilson 1988).

Conclusion

We began this chapter by asking whether NC/CNC – the most successful, independent, technological development in metalworking in recent years – was a continuation of a Taylorist logic of hierarchical controls and deskilling, or whether it was the unwitting breakthrough into post-Fordist manufacturing posited by Piore and Sabel? Answering this question required consideration of further problems: why did NC succeed, against what alternatives, and for what reasons – technological or socio-political?

So far we have seen that the success of NC/CNC rather than the alternative approaches of GT and integrated cells arose from a mixture of *both* technical factors and socio-political institutions. If we assume that the general and prevailing social relations in capitalist enterprises are more or less given, then technical and general economic factors seem almost sufficient to explain NC adoption and the patterns of its use. Almost, but not quite. The success of NC lay in two advantages perceived by production managers. Firstly, it allowed a continuation of the piecemeal rationalisation of the Taylorist paradigm, without the more fundamental attitude shifts and process reorganisations required by the quasi-Fordist GT and integrated cell approaches. Secondly, it allowed managerial discretion in the use of labour skills. If appropriate, NC's original potential for hierarchical control, and diminution of operator discretion, could be pursued. However, if this was difficult, or impractical, then skilled labour could be retained, albeit with a modified repertoire of skilled tasks.

However, the skill compatibility of NC utilisation, and hence the success of this technology, cannot be restricted solely to the technical practicalities of batch work. To restrict the cause of the exercise and deployment of skills to these factors would make interpretation of socio-industrial change a highly predictable and formalistic activity. The chronological and organisational diversity in skill applications suggests this is not plausible. At the extremes most influence may come from the logic of either very small or large batch sizes, of high-quality or low-quality materials, and from the necessary precision, of either simple and infrequent, or frequent and complex, resetting. All of these factors will cause either further deskilling of machine operation or the retention of skilled operators.

On the other hand, as argued elsewhere (Braverman 1974; Noble 1984; Jones 1983), one can hypothesise a work organisation where the social relationships and economic constraints specific to contemporary industrial capitalism did not exist. In this world there could be a logical organisation of work where it would be technically possible for one type of worker to carry out all the varied tasks of computerised machining. So socio-economic relationships could transform the division of labour and hence the use of the technology. To discover the precise ways that the intermediate sphere of skills is shaped by complexes of social conditions in our present reality, we can simply ask whether skills are used in a more flexible union with the computerised technology, or whether the latter is used as a more rigid – though bureaucratically 'efficient' – extension of Tayloristic organisation?[11]

In particular, how is the definition and availability of skills embedded in a variety of actual practices? This is where the role of socio-political institutions has also to be considered. Actual practices are themselves inherited from bureaucratic organisation, managerial paradigms and prerogatives, occupational specialisations, labour unions, and the industrial relations institutions – all of which mediate between workers, unions and managers. Skills and tasks are made available, and managerial choice of practical combination with NC systems is made possible, because of the influence of these social institutions. The 'embeddedness' of practical skills in these institutions appears as the ultimate arbiter of the technological selection and economic viability of NC against rival schemes. Ultimately, the deciding factor appears to be the mutual compatibility of the machines and the social-institutional 'environment'.

It is for this reason that we must now consider the different patterns of NC/CNC usage in different firms within separate, national institutional contexts. The answer to our initial question was that there was, as labour control perspectives argue, some continuity with Taylorist logic in the early phases of the development and take-up of NC. However, this consistency broke down when firms with varying product characteristics, or small batch sizes retained skilled operators, or setter/operators. Later, when CNC provided the capacity for shopfloor programming – or, at least, program modification – use and adaptation of craft-type skills was facilitated; especially in smaller firms. This last development is part of the Piore and Sabel thesis of a discontinuous shift back towards craft skills as a consequence of the Pandora's box qualities of CNC. However, as Piore and Sabel acknowledge, the scope for national variations is also significant.

The conceptualisation of skills in this chapter, as the bridge between the technical and socio-political, suggests that national institutions are crucial in defining and providing skill resources in firms. The patterns in which skills have been combined with NC/CNC are also important for the subsequent applicability of more Fordist forms of cybernation. Does the happy marriage of craft skills and CNC rule out the neo-Fordist appeal of more integrated forms of computerised production? Or have managers continuing ambitions for greater control or productivity? In which case the form and extent of the persistence of skilled labour may either predispose or prejudice managers towards the promise of cybernated factories. We need to assess, therefore, in some detail whether, and how far, these mutual adjustments of NC/CNC to craft skills took root in specific national contexts.

5 Numerical control, work organisation and societal institutions

Plant managers don't choose between Fordism, Taylorism and their successors. For some academic analysts technological change means managements, especially 'successful' managements, making explicit decision-making reviews of the relevant aspects of their operations. Rational choices are made according to the goals the firm as a collective manager seeks to achieve. Outsiders may criticise the scope of these goals, but within these limitations it is assumed that deliberate decisions are made (cf. Bessant et al. 1992). The more articulate 'strategic choice' theories of this kind have been criticised for their failure to take account of the ways in which differences in participants' organisational power and normative outlooks restrict and shape decisions (Thomas 1994, pp. 213–31). In the case of NC/CNC it seems that participants' paradigms rarely 'saw' the implementation as requiring decisions on reorganisation. Often they were not consciously deciding, but merely following – from current paradigms – the habitual practices defined by national institutions for labour management, industrial relations and occupational responsibilities.

In theory, those who adopted NC or CNC technology encountered a myriad of potential and potent decisions about organising and staffing. Questions about departmental responsibility of programming and planning, about management control of the supply and monitoring of parts and tooling, about locating the new machines amongst existing machining sections or in their own special areas, and whether existing shopfloor managers, or a new category of NC supervisors should supervise. An especially significant issue from our point of view concerned work organisation. Should NC's deskilling potential be realised by drafting in unskilled or semi-skilled operators and leaving the programming completely to the part programmers? What role should be accorded to the shopfloor workers needed for the remaining skilled tasks of setting and maintenance?

In practice many decisions on these issues were never consciously taken, or never emerged as clear-cut and manifest choices. Large firms, such as major aerospace contractors, could establish entirely new machine shops and planning departments as a result of their simultaneous investment in scores of new NC machine-tools. Smaller firms, or those with less project-based investment strategies, might purchase NC and CNC incrementally in ones and twos. In which case even thinking about fundamental reorganisations was *a priori* uneconomic in time and money. Elsewhere however, in between these extremes, managers did decide – if only by default – between at least one of the above pairs of choices. In some areas managerial prerogative was unfettered; in assignment of organisational responsibilities or locational decisions, for instance. Then only intra-managerial politics complicated the decisions. But where the division and allocation of skilled tasks was concerned decisions were heavily influenced, in some instances pre-empted, by the weight of the social institutions in which skills are defined and organised.

Centralisation and specialisation of the computing aspects of the new operations would not only maintain the continuing viability of existing paradigms but influence attitudes and practices for the future stages of further computerisation. Involvement of operators, and/or craftworkers, on the other hand, would contradict Taylorist norms and widen the range of future options. Lurking behind these possibilities was the continuing tension between factory and workshop modes of organisation. The tasks of this chapter are therefore both descriptive and analytical. We need to return to the socio-political trajectories, especially their industrial relations aspects that we left in chapter 3, and to review what is known of their interaction with the adoption of NC technology. Analytically, it is necessary to underpin this description with an assessment of the relative influence of particular, technical and organisational factors and the national institutions which shape choices over the balance of skills and technology.

To demonstrate the extent and relevance of these different causes and conditions requires, in the first section below, an assessment of the influence of administrative organisation. A significant tradition within managerial and organisational studies extends back, in a prescriptive sense, to Taylor himself and analytically to the work of Woodward and the Aston school. This approach accords primary, and potentially universal, influence to the type of organisational structure (Woodward 1980; Brown 1992; Rose 1988; Child and Mansfield 1972). I have already suggested that organisational character is a plausible cause of skill utilisation: where bureaucratic processes and hierarchical decision-making make possible, and also co-ordinate highly demarcated job specialisations. By contrast, in smaller plants the tendency is for less specialisation and more spontaneous decision-making.

There is one further complication when assessing the interaction of technological development with social institutions and interests: the latter do not remain constant. These, too, are in a continuous state of greater or lesser change. Therefore, the second section of this chapter outlines the relevant industrial relations and plant politics. It also shows that the amount of change in these social institutions, during the NC growth phase, varied considerably between countries. The overall framework in the United States remained largely static while the Italian institutions underwent radical reforms. Hence the social influences have to be treated, not as fixed variables, but as dynamic processes whose trajectories of development may impinge in different ways on the skill utilisation decisions in the firms.

Causes and conditions

Our heuristic contrast between the deterministic deskilling thesis of Noble and the contingent pluralism of Sabel shows the simultaneous absence and importance of social institutions.[1] If there is deskilling, it is not machines who sack skilled workers. If particular skills optimise new technology, then those skills have to be supplied and supported by appropriate institutions and organisations. Three types of institution help to define, allocate and exercise NC skills. One is the firm's distribution of skills into occupations and functions; effected through its training decisions, recruitment practices and departmental responsibilities, but also socially conditioned by national training services, qualification arrangements and occupational specialisms. The second institution is the administrative organisation of production operations, most visible when larger factories segregate the various operative, maintenance, supervisory, service and planning functions into specialist occupational positions. The third factor is the 'politics of the plant' (Rose and Jones 1985): the overt industrial relations (IR) system, and also divisions and conflicts of interest amongst managers and amongst workers. Such factory-level politics are localised in their outcomes but national IR institutions and politics set their broad limits.

Hence when some British managers 'deskilled' jobs of NC operators this was only politically feasible because of temporary weaknesses or bargaining tactics of the main metalworkers' union, the AUEW. Moreover, such a deskilling of operators' jobs was only practically feasible because the semi-skilled operators could be supported by the reserves of craft-skilled manual workers. These were available for setting-up tooling and workpieces and 'trouble-shooting' mechanical problems. Or, to take another example, firms' belated recognition that accurate part-programming requires varying degrees of machinists' craft skills could only be put into practice if programming staff could be recruited from shopfloor workers; or if it could be shared between the programmers and the setters. The capacity to choose one of these compensatory skill arrangements depended on organisational and industrial relations rules governing the transfer, or promotion of workers, and the demarcation of task responsibilities between different categories of worker.

Assessment of the causes of skill deployment is also bedevilled by assumptions that there is only one absolute, and ultimately inescapable, method of operating efficiency for a given production process. Both the deskilling and reskilling scenarios imply that there is one best way to operate NC. For the former that arrangement is maximisation of managerial control and 'top-down' achievement of prescribed standards by Taylorist bureaucracy. For the reskilling scenario, in the craft paradigm mooted by Sabel et al., decentralisation of part-programming and setting responsibilities to shopfloor workers, or collegial work groups, is the optimal arrangement. Such absolutist characterisations do not accept fully that either extreme, or a number of intermediate arrangements, may be satisfactory. If the managements involved see the relevant efficiencies of either mode as being paramount, and the costs of alternatives as being punitive, then either of these two approaches may seem effective and 'successful' to them. If a pattern does emerge then rational choice by managers is probably guided by broader institutional conditions. To see why one emphasis rather than another comes to be generally preferred we must analyse the extent of this interplay between local decisions and the social institutions which shape attitudes and hence organisational choices.

Figure 5.1 suggests two analytical types of NC operation and skill utilisation. The first mode aims at hierarchical controls and minimal

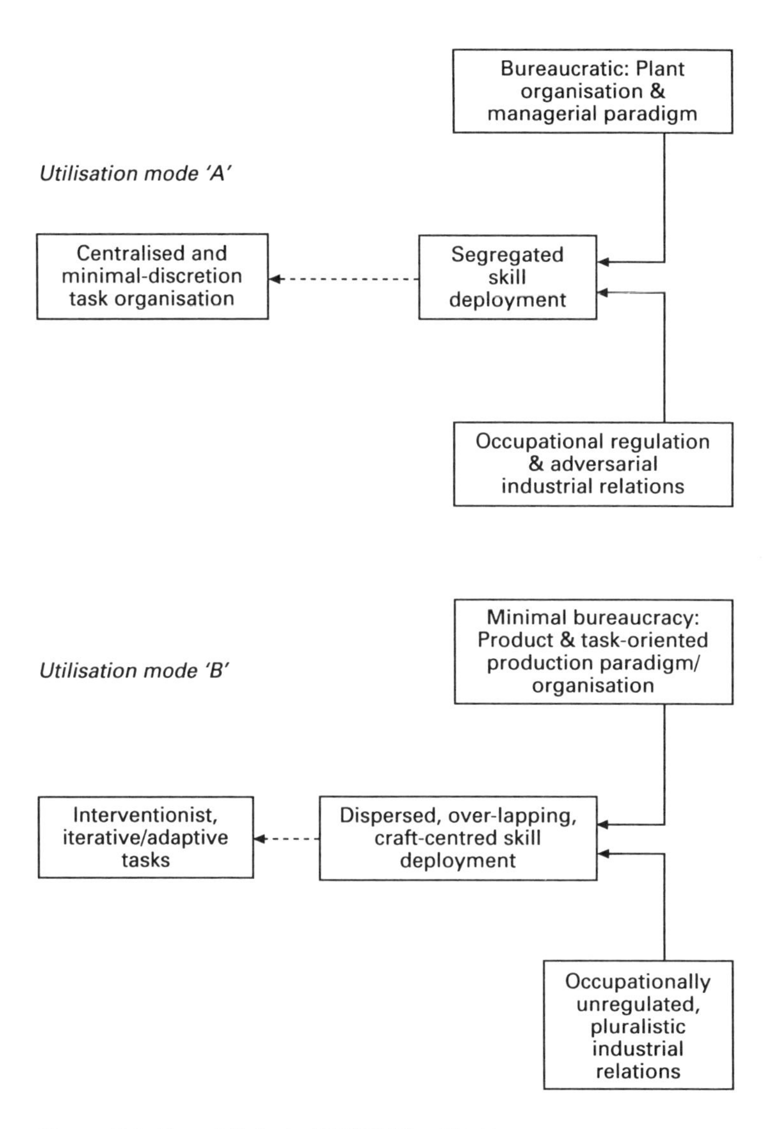

Figure 5.1 Causal links in NC/CNC utilisation

modification of operations and programs. The most logical skill distribution accompanying this mode is a centralised monopoly of skills, mostly in the planning–programming departments. Tasks are sharply segregated, at least between the shopfloor and the planning/programming functions, possibly with tight occupational boundaries between all jobs. The most likely institutional influences on this arrangement are a deeply ingrained bureaucratic culture and plant organisation, and industrial relations characterised by regulation of occupational specialisms and an adversarial culture and history.

In the second mode an adaptive style of using programs with continuous minor refinements on the shopfloor is achieved by a spread of craft-based skills which tend to overlap a range of different operating, setting, maintenance and programming occupations. Accompanying this 'fuzzy' task organisation is an adminstrative structure with minimal bureaucracy and a fluid set of operating objectives and practices; something approaching the famous organic form of organisation (Burns and Stalker 1961). Industrial relations, insofar as they influence the running of the plant, will place a low emphasis on occupational specialisation and regulation, and tend to be pluralistic in character. Both of these modes are abstract types. For in any one society, sector or factory, it is likely to be a combination of operating style, skill deployment, administrative organisation and industrial relations institutions – operational and institutional influences – which will determine outcomes. A legion of research has established the specific character of these local organisational forces as major influences on the precise division of skills and labour between occupations and departments (see Thomas 1994, p. 22). Yet the relevance and strength of intra-organisational influences varies between countries. Maurice and his colleagues have shown emphatically the force of these 'societal effects' of national institutions (Maurice et al. 1986). So the effect of the societal context of the local–organisational dimension has also to be assessed.

Administrative organisation

In a recent survey of all forms of 'Programmable Automation' – but mainly NC/CNC applications – in a sample of US metalworking plants, Kelley took plant size as a proxy for bureaucratic organisation. In these workplaces where there were 1,000-plus employees – and where other factors such as unionisation and batch size were taken out of the analysis – programming responsibilities were 50 per cent more likely to be 'exclusively a white-collar job responsibility'. However, at the other end of the size spectrum, where bureaucratic-Taylorist organisation is much less likely, the situation is reversed. Machine operators are three times more likely to do programming than other occupations (Kelley 1989, p. 242). An apparent European confirmation of this organisational influence comes from the case study comparisons of British and West German CNC users by Sorge et al. Consistent with Kelley's evidence large firms were less likely than smaller ones to allow machinists a role in programming.

Moreover, exceptions to such a division occurred in the German rather than the British plants (Sorge et al. 1983, pp. 83–97) – no doubt reflecting German emphasis on craft skills on the shopfloor, which is less conducive to Tayloristic work organisation (Lane 1988, pp. 143–8).

Similarly, Dodgson's study of British small-firm users of CNC found a difference between completely independent enterprises and those financially or administratively controlled by a larger organisation. More 'polyvalent' machinists – those who also set up and, at least partly, programmed the machines – were more common in the independent firms (Dodgson 1985, p. 81). Mere smallness in size – measured in numbers employed – was more likely to be associated with a non-bureacratic organisational character (Dodgson 1985, p. 67); which Dodgson also linked to a distinct 'managerial ethos'. However, there was no necessary correlation between size and work structures, and organisational character was partly independent of size. This finding was supported by another analysis of a small specialist CNC jobbing-shop, where the dominance of planner-programmers left no programming tasks for operators (Jones 1982). So bureaucratic organisation is a more important cause of the fragmentation of tasks and skills rather than size as such. But why is this so; and is it the most effective administrative framework for organising CNC operations?

Both Taylorism and Fordism, concentrate planning and design tasks into specialist occupational functions. These sources attempt to set standards in methods and components and hence give rise to contemporary industrial bureaucracy in small-batch manufacturing. One could see these developments, as in the labour process perspective, as part of a continual perfection of bureaucratic control through increasing fragmentation of technical knowledge: firstly by transferring the 'conception' tasks from the craftworker, and then by dispersing these responsibilities amongst more and more detailed clerical and technical occupations (cf. Braverman 1974, pp. 239–47). However, it is far more difficult to prove that actual operations on these lines do achieve the reputed efficiencies. More focused studies have revealed the contradictions within the bureaucratic processes by which scientific management administers and co-ordinates specialised skills. In particular, the processes needed to link all this fragmented knowledge into the operations for a finished product, may themselves be a drag upon any general deskilling.

Conflicts and disputes over product specifications between design engineers and production staff – leading to departures from standards – were diagnosed as symptoms of bureaucratic 'mechanistic' organisation structures in Burns and Stalker's classic study of organic and mechanistic administrative structures (Burns and Stalker 1961). In a detailed study of a large aerospace plant, Smith recorded discrepancies between the standards set by designers and those expected by the detail draughters responsible for the next stage of product and component specification. Inevitably there were also mismatches between the specifications encoded by the part-programmers and those expected by the machine operators (Smith 1987, pp. 145, 219–20). At another British machine-tool manufacturer the new programming technicians contested the sphere of responsibilities with a longer established section of planners; and the latter group's competence was frequently challenged by the shopfloor workers (Wilkinson 1982, pp. 59–60).

A particularly vivid illustration of the contradictions between occupational specialisation and the accompanying bureaucratic co-ordination was found in a large NC operation in 1978. Operators were making repeated queries and referrals back to part-programmers because rules forbade operators to modify programs themselves; and the control panels on CNC machines were locked. Yet the machines would often lie idle until the problem had been resolved. Operators had got into the habit of telephoning for the relevant programmer to come to the shopfloor; or even, on some occasions, going into the programming office themselves to discuss the problem with the technician who had written the program. This *ad hoc* practice upset the programmers' manager because it distracted them from completing other part-programs. Authority prevailed and a formal process was instituted whereby problems and queries were written out by operators, passed to machine shop supervisors who – if they agreed with the operator's diagnosis – would then pass the note to the head of programming. If remedial work would not interrupt the completion of new programs, the latter would then authorise the part-programmer to spend time resolving the machining problem. This was a rational solution from the point of view of the programming department logistics; but costly NC machine-tools stood idle for as long as a week while the paper requests travelled up and down the hierarchies (Jones 1982, p. 192).

More generally, rigid occupational demarcations between NC task skills tend to complicate the completion of programs and remedial actions to optimise them (Jones 1983). Specialised occupational roles and bureaucratic regulation of the technicalities and economics of production all dictate that tasks must be divided between specific categories of worker. They are very real social constraints. These different interest groups, and their politics, together with those of management, and of the formal institutions of industrial relations, make up the final determinant deciding the division of skills and responsibilities. They thus control the extent of deskilling, or reskilling. They also, as a consequence, limit the effectivity of the new technology as a productive force. However, the presence and strength of some of these factors is not universal. They are specific to a culture or society. So national differences need also to be identified.

NC established itself in large plants and technologically advanced industries because it fitted into existing managerial paradigms of hierarchic task controls and bureaucratic procedures. In contrast, the localised and more flexible programming traits of CNC later matched the more decentralised and *ad hoc* decision-making of smaller organisations. In the first type of firm labour and attitudinal factors mitigated the numerous operational problems arising from this use of NC under centralised control. This was partly through the existing stock of skilled manual workers in these plants. It was also partly because managements often saw inadequacies as temporary costs in a much longer-run automation strategy. In this latter view, operational shortcomings would be transcended as the firm adopted more advanced forms of computerisation in the future. For the larger European and North American firms, we must ask why, and how far, socio-political institutions of labour regulation (Batstone et al. 1987), broader than the organisational factors discussed above, shaped the skill inputs which, in turn, influenced NC usage?

Plant politics and national industrial relations

Chapter 3 described the interaction of industrial relations trajectories with post-Second World War rationalisation of production jobs through Fordist standardisation and Taylorist job design. We left that account after assessing how far, and with what success, unionised labour in the three countries responded to managerial initiatives to rationalise production. We can now extend this comparative description of trade union influence to the organisation of work with NC and CNC in the 1960s and 1970s. British unionised workers responded through the perspective of *craft union job controls* (cf. Zeitlin 1979). In the USA the framework of union action could be described as *contractual individualism*, and in Italy unionised workers were most influenced by a political-cultural movement for *egalitarian solidarity*. The meaning and significance of these labels will be brought out by describing developments in each of these countries in turn.

Britain

Here the main union concerned has been the craft-based Engineering Section of the confederal Amalgamated Union of Engineering Workers –

since de-merged as the Amalgamated Engineering Union, and then combined with the electrical and plumbing workers union to form the Amalgamated Electrical and Engineering Union. The AUEW largely left bargaining over new technology to local, and especially workplace representatives: 'shop stewards'.[2] These representatives, and often the workers they represented, were mainly concerned with the impact of technology on the numbers of members' jobs and on maintaining the skilled *status* of those jobs. This last point needs emphasising. Workers and individual shop stewards were often concerned about the objective loss of the exercise of more traditional manual and control skills through technologies such as NC. However, they had inherited a union tactic for coping with technological and organisational change which was directed at preserving the social status of the craft – i.e. apprentice-trained – worker.

Craft-union policies

This tactic, reportedly stemming originally from the acute labour shortages during and just after the Second World War (Jefferys 1945, pp. 245–6), involved negotiating the 'dilution' of jobs previously regarded as the responsibility of craft-trained workers. The union would agree to managers installing employees without craft status, subject to certain provisos negotiated by union representatives. These included mutual recognition that a job that was 'diluted', because of insufficient qualified workers, could, in principle, be reclaimed for a craftworker, if one became available. Moreover, the union would always aim to keep the rate of pay for the non-qualified 'dilutee' at, or about that agreed for craft grades.

With the introduction of NC equipment, and management claims that machine operation was now a less skilled job, the union response was to argue that the greater sophistication and cost of the new machines required higher levels of responsibility, and thus skill, from operators. This kind of counter-argument was difficult for managers to rebut when it was accompanied by experience of the kind of costly failures in part-programs, described above. Most of the British studies record unions using these and similar arguments to resist the operation of more than one NC/CNC machine by the same operator. Craft sentiments were rallied with the slogan: 'One man – one machine', even though on long machining cycles, on well-proven operations, operators would frequently complain of boredom and idleness. If necessary union bargainers would fall back on dilution provisions: 'yes' – NC operator jobs could be taken on by semi-skilled workers, providing they received the craft grade wages, and providing the jobs could – at some specified point – be reconsidered for occupancy by craftworkers. Such conditions would most likely mean that managers could not expect significant cost savings from dilution. Unless there was an absolute scarcity of craft qualified workers it was almost as cost effective to retain skilled operators as it would be to replace them with semi-skilled ones.

National AEU policy has exhorted local representatives to press for operator programming of CNC since 1981.[3] However, the ambiguous wording of this Executive Committee recommendation argued that AUEW 'craftsmen' should apply, and be eligible for, any vacancies advertised for CNC programmers within a particular company. This implied that programming would continue to be done away from the shopfloor by promoting craftworkers into those office jobs. Yet the circular also argued that negotiations should always press for 'proper training for all grades of employee involved in CNC operation'. It also said 'Programming can logically be regarded as the logical extension of setting up tools and machines, and it is *therefore* rightly the province of our members'(original emphasis).

Workplace politics

However, explicit plant-level union action to gain the right to shopfloor programming or editing has rarely been recorded. Any such arrangements for operators and setter-operators have been gained largely because of a lack of detailed policy by senior management. Shopfloor programming seems more often to have arisen through individual struggles and disputes between machinists and programming and planning staff. Yet a strong, indirect, union influence has been involved: the continuing culture and traditions of shopfloor independence, and the tacit and sometimes open support of shop stewards (Wilkinson 1982, pp. 63–6; Jones 1982).

Where the shopfloor ethos and production technicalities have been favourable, union representatives have taken up shopfloor programming as an explicit aim (Jones 1982; Batstone et al. 1987, pp. 141–55). However, for success through such more organised and strategic action more complex pre-conditions need to be present. For example, the small-batch components plant in the Batstone study was like those discussed in the Taylorisation campaigns in chapter 3. It had absorbed scientific management techniques into the older British style of craft administration. So the union, and even individual workers, negotiated on individual rates and allowances for jobs performed on the PBR scheme; and there was no work study department. Like British managements' response to scientific management earlier this century, key management figures in Batstone's study positively preferred this element of craft administration to the bureaucracy of Taylorism. They were sympathetic to the continuation of craft skills and independence, seeing them as important motivators for individual operators' performance in making small batches of parts to high levels of precision.

A more significant political issue in this case related to the administrative organisation of production. Technicians had only a limited occupational identity. Elsewhere, these might assert their own interest in controlling programming functions. But in the Batstone study a specialist programming role for technicians in the planning area was never established. This may have been linked to divisions in the union organising the technicians. Unlike the USA or Italy, in larger British manufacturing firms it is common for the latter to belong to a union. Until recent union reorganisations this was usually TASS – the Technical, Administrative and Supervisory Section of the AUEW. Despite its official affiliation with the manual workers' sections in the AUEW confederation, effective co-operation between TASS and AEUW was rare. Batstone et al. found divisions between the planning staff and the supervisors over formal responsibility for CNC. The latter effectively sided with the craftworkers against the planners (Batstone et al. 1987, p. 140). Similar pre-conditions favouring operator involvement in programming occurred elsewhere (Jones 1982). Here a strong and strategically minded AUEW faced a weaker TASS organisation. Management wanted to focus programmers and planners on basic design and work processing issues rather than the detail of part-programming.

Thus there have been strong industrial relations pressures favouring involvement by NC/CNC machine operators in the programming, or program modification, tasks in British factories. Workers, stewards and union organisations have capitalised on the technical and practical advantages of operator involvement. The operator is available and knowledgeable about the mechanical characteristics of particular machines and machining processes. Whereas programmers, even those who have previously worked on the machines, may not have detailed and recent experience; and cannot be readily consulted on the shopfloor. In many plants these practical advantages have been given extra salience by the impact of union bargaining. This has not always aimed directly at retaining the *exercise* of craft-type skills. Yet it has often *indirectly* had this effect by limiting the management's financial advantages in employing less skilled labour. A measure of the strength of these combined social and practical pressures is indicated in a 1982 AUEW survey of NC/CNC staffing arrangements. This showed that by 1982 programming was the responsibility of AUEW members – effectively setters and operators – in just over half of the cases reported in the union's various regional divisions.[4]

The United States

In the USA the legalism of the collective bargaining process, and the attendant individualistic focus, in translating workers' grievances into checks on the potential deskilling of jobs through NC, warrants the term 'contractual individualism'.

The pattern of worker involvement

Despite some superficial similarities the involvement of shopfloor workers in programming tasks is much less pronounced in the USA than in Britain. In her survey of all forms of programmable automation – that is NC, CNC and the more complex Flexible Manufacturing Systems (FMS) – Kelley found that shopfloor workers had complete programming responsibilities in only 13.1 per cent of plants (Kelley 1989, p. 239). Although 44.4 per cent reported that programming was 'shared', this probably means that setters and operators modified elements of the programs or partly refined them at the prove-out stage. So the ratio of shopfloor programming seems significantly smaller than in Britain.[5] Other evidence suggests that in the American situation the industrial relations factor has, unlike Britain, *weakened* the practical and organisational influences favouring shopfloor programming.

Industrial relations institutions

Much metalworking manufacturing in the USA is located in the industrial north and north east. Here, for historical reasons – at least amongst larger establishments – union organisation has become focused on the local plant. Union success in raising hourly wage rates has led managers to invest in new capital equipment mainly to minimise inputs from unionised labour (Kelley 1984, p. 24). But why should this be different from the British situation? In Britain union density in these sectors is also high and, as we have seen, union pressures can retain a strong role for skilled production workers on NC/CNC. The difference lies in the institutional character of labour relations in the USA. For legal and historical reasons union membership rarely covers professional and managerial workers. Consequently, classifying the planning and programming of NC/CNC machines as 'non-manual' gives much more freedom of manoeuvre to managers wanting to maximise control over the amounts and costs of labour. By declaring that programming duties have a supervisory element managers can appeal to legal processes, if necessary, to exclude them from union jurisdiction (cf. Jones 1984).

The differences between larger unionised plants and larger non-union plants in Kelley's survey confirm this restriction. For every single case of the former, where shopfloor staff had some involvement in programming there were – depending on plant size – between two and four cases with a strict Taylorist separation of programming tasks from machining work. By contrast, in the latter, non-union, type of plant, for every one case of strict separation of programming and operating there were 1.6 contrary cases of some shopfloor involvement (Kelley 1989, p. 244). Demarcation of labour and management spheres of prerogative in unionised US plants is thus a decisive factor which segregates programming from operating tasks. The more advanced technology of Flexible Manufacturing Systems, described in the next chapter, shows that the mere absence of unions does not make managers willing to combine operating and programming tasks. Historically, however, because the roles of labour, management and the state have been brought within legally regulated institutions this has probably been the single most pervasive influence.[6]

In the USA unions have had the right to bargain over *who* should be retained or allocated to new or changed job categories. But it is management's right to specify in detail *what* work is to be undertaken. American labour law is based upon the principle of separate and self-contained rights. Where the union believes that these rights are in conflict, and management disagrees, then reference must be made to the legal procedures of the national labour relations courts. Unions know, however, that any retaliatory action to pressure management on a job classification could mean that management could claim to the courts that the union had broken its contractual agreement with management. Exemplary cases are the experiences of the International Union of Electrical Workers (IUE). Because of provisions for only single-union representation in the legislation, this union represents many machine-tool setters and operators in companies such as the giant General Electric (GE) corporation.

Union tactics towards NC

With the advent of NC the IUE sought to have the operation classified as equivalent in skill to comparable work with conventional machines. This action entailed detailed submissions and hearings on individual test cases in the tribunals of the National Labour Relations Board. The unions eventually lost these cases. In more recent years some US unions seem to have succeeded in getting local managements to rate operating NC machines as comparable in skill, and hence pay levels, to the operation of conventional machines. However, this was a more arduous contest than that of British unionists. So unlike the British case, managers could, if they chose, claim that the determination of the task content of a job was a managerial prerogative.

Henceforth, the IUE and other US unions, such as the International Association of Machinists, have been constrained to pursue more *ad hoc* and indirect tactics. Union locals have, for example, used analogous tactics to the British unionists' dilution manoeuvre. They assert the seniority principle which allows priority in access to new jobs to more experienced workers. Thus, it could be claimed that a skilled machinist, due to be displaced by NC, was the most senior candidate for the new NC operator. Management could then be pressured to maintain wage levels when the operator showed that he or she was still exercising important skills (Jones 1984). But managements were resistant to an 'upward' extension of operators' skills into programming tasks.

Local management at GE's Lynn plant in Massachusetts, recognised in the late 1960s that considerable skill and involvement was needed to operate and adjust NC machines. Their concern led to the formation of a special pilot group with interchangeable operating and maintenance roles. These eventually became self-monitoring and responsible, also, for modifications to the part-programs. When the IUE sought to extend this experiment to other GE plants, and the developments at Lynn were reported in a national newspaper, GE's corporate management claimed ignorance of the local managers' reforms. Then they had the Lynn pilot group itself abolished. Joel Fadem's account – unlike that of Noble (Noble 1984, pp. 265–323) – suggests that one of the key issues that disturbed GE's corporate management was that 'more fluid work groups might not conform to National Labour Relations Board definitions of exempted jobs carrying management responsibilities' (Fadem 1976, p. 14). In other words, NC planning and programming work could fall into the sphere of union influence if it were not rigidly separated into managerial positions.

So the US experience has been that the influence of unions and industrial relations institutions was both less equivocal and more tortuous than in Britain. US industrial relations struggles through 'contractual individualism' have been a process more bureaucratic and probably less influential than the British experience of collective craft controls.

Italy: the general pattern

Italy presents a more confused picture than the other countries. This is partly because of the shortage of detailed studies. A more basic factor is that the industrial relations themselves became more complex and volatile during the 1960s and 1970s. Yet the 1970s were the decisive growth phase of NC/CNC investment. Therefore, as far as work organisation is concerned we have to generalise retrospectively from a small number of significant cases. So it has to be admitted that the evidence for making any judgement is limited. Nevertheless it seems the new forms of industrial relations prevented the wide Tayloristic centralisation of programming and control more typical of the US case (cf. Kern and Pichierri 1996).

Numbers of such machines grew massively from a mere 400 in 1968, to 4,400 by 1975, and again to 11,600 by 1983 (Dina 1986). On an apparently more generous definition Italy, with 82,566 NC/CNC by 1985, had a higher proportion than Japan, or other West European countries (*Metalworking Production* 1988, p. 15). During the first of these phases the pendulum of shopfloor power moved towards workplace unionism, away from the employers; and, in some respects, away from the central trade union organisations. A heterogeneous and unstable pattern of workers' influence on NC usage arose from this conjunction of rapid adoption of NC technologies with wholesale shifts in both the balance of industrial relations power and its institutions. Italian production workers gained involvement in preparation and programming, but not to the extent suggested by their influence and militancy in the early 1970s. In aggregate terms the main outcome appears very similar to the eventual British pattern. However, the causes of this involvement arose from different union philosophies which contrast sharply with the Anglo-American histories. As we saw in chapter 3, in the immediate post-Second World War decades the reconstruction and stabilisation of Italian industry was shaped by a political environment in which unionism was divided, discouraged and even persecuted. This weakness facilitated employers' imposition of the crudest types of Taylorist exploitation. Ironically these very abuses stimulated the collective grievances which provided the springboard for the massive worker protests that accompanied the final phase of post-war expansion in the late 1960s.

Industrial relations as social transformation

This protest movement, started by traditionally unionised workers over conventional economic issues, grew during 1969 and 1970 to encompass firms, sectors, regions and occupational groups previously peripheral to organised trade unionism (Regalia, Regini and Reyneri 1978). In the industrial sectors its motive force was long-standing grievances over low wages, long hours and conditions of work imposed by the 'ragamuffin Taylorism' of the early post-war period. But its ideology and main constituency was much more radical in scope and potential. Della Rocca argues that the very new and radical unionism of the late 1960s stemmed from a combination of authoritarian Taylorism, the compromised position of the Left unions – who supported labour intensification in the immediate post-war 'reconstruction' phase – plus the elimination of the older union ties by the employers' offensive (Della Rocca 1976). The 1960s produced reactions against the divisiveness caused by managerial attempts to further the job evaluation schemes first imported from the USA under the Marshall Plan; reactions which provided another impetus (Barisi 1980, p. 11).

The primary focus became the semi-skilled workforce and its social, economic and political position in the enterprise. Influenced by a more radical Marxism than that of the Italian Communist Party, the dominant rationale of the protest movement was made up of three key ideas: economic equalisation; political mobilisation around the 'mass worker'; and the abolition of specialised work roles.[7]

'Economic equalisation' meant greater solidarity in wage bargaining: reducing, or abolishing the multiple distinctions of incentive payments, job evaluation schemes and differential wage increases. Together these constraints had isolated large numbers of the unskilled and semi-skilled workers. The 'mass worker' concept adapted a more potent Marxist concept of political change. The self-mobilisation of the homogeneous mass of factory operatives would form the critical political force for social transformation. Their social and economic position, it was held, was a concentration and a pivot of all the key divisions of authority, power and exploitation in society at large. Therefore the emancipation of the mass worker could also set in train a broader social revolution.

The metalworkers' unions were a central force in the Italian labour movement and in the 1960s these industries expanded employment rapidly – conditions which encouraged the Promethean image of a politically crucial force of rank-and-file factory workers. Despite their separate policy sources, economic equalisation and emancipation of the mass worker had similar influences on work organisation. The theme of transcendence of specialised work roles combined elements of equalisation of pay and the anti-Taylorism in the mass worker line. Levelling wage differences, and abolition of divisive incentive payments, had ideological resonance with the overthrow of political and social subordination. These aims had separate sources and advocates, but a common theme: the challenge to managerial definitions of task, occupational and status divisions within the factory. For some campaigners, schemes which expanded work roles for higher skill status would eliminate the worst inequalities and set examples for society as a whole through greater autonomy, discretion, and range of skills for ordinary production workers.[8]

It should be noted that this ethos was neither homogeneous nor universal. It may not have been operationalised in many instances. The social philosophy of Catholicism, influential in some union circles, had an ethic for the humanisation of factory work. However, its main influence was in the equalisation of pay. Promotion of new forms of work organisation, on the other hand, came not from the Catholic/Christian Democrat-aligned CISL (whose sectoral organisation is the *Federazione Italiana Metalmeccanici*), but from the communist CGIL. More pragmatic calculations also, no doubt, guided grassroots tactics. Despite these complications, and the fact that equalisation was only an official policy with respect to wage reform, a diffuse culture of egalitarianism linked the trinity of monetary, political and work reforms. It also articulated workplace militants and their supporters to the national union organisations throughout Italian industry (Heine 1986).

It is significant that the skilled workers, those most likely to be affected by NC/CNC installations, were central figues in the struggles against Taylorism in the early 1960s. Yet in 1969 the metalworking unions started the so-called *professionalità* movement. During the early 1970s this campaign aimed to complement pay and work organisation reforms with skill enhancement through requalification for semi-skilled and unskilled workers, rather than the craftworkers, who had predominated in campaigns prior to the militancy and egalitarianism of the late 1960s (Barisi 1980, p. 10). This overshadowing of the skilled workers should not be exaggerated. As we shall see in the next section, they were able to promote their own interests – in particular plants, and in matters such as the control of NC machinery – by varying the egalitarian line to oppose the role of technicians.

However, in general, this politico-cultural movement had several implications for organising work with the concurrently spreading NC technology. Probably the most important consideration was the sheer diffuseness of the egalitarian philosophy. In the USA work reform was absent from trade union discourse. In Britain it was translated into the traditions of militant, but pragmatic, craft-unionism. In Italy, however, the relationship between work reorganisation and the trade union was: 'not just the starting point of a more far reaching critique but a central theme of all its activities' (Rollier 1976, p. 83).

Thus union bargainers could hardly fail to see the implications of NC working practices for the division of labour and the quality of work life. Yet the philosophy of emancipating the mass worker through expanded work roles could mean various constraints on the scope of union tactics towards NC work. First, and most obviously, egalitarian policies were oriented primarily to other aspects of semi-skilled production work. For example, raising skill levels in routine jobs such as component assembly, would have higher priority in multi-function plants. Such emphases would overshadow the defence of existing skills, such as the more responsible machine-setting and operating in small-batch production.

Secondly, the preferred changes for enhancing work were also more geared to the semi-skilled tasks, rather than to the still relatively varied role of the machine-tool setter, or small-batch operator. Policies for job rotation and to abolish pay-for-output schemes – on the grounds that the individual worker could not greatly influence output – would have less salience for workers on machine-tools with frequent resetting. Some rotation is usually part of their training requirements, and is often implicitly expected to cover the fluctuations in machining processes that are inherent in small-batch production. Furthermore, as Burawoy and the classic US studies show, machinists can realistically expect to exert an influence over piece-work that is denied to production-line workers. Moreover, because NC predetermines many operating times, and can be used to maintain pre-set speed and feed rates, it frequently makes such incentive schemes inoperable anyway. It was a curious paradox in some British shops that workers and unions opposed NC because it meant the end of piece-work payment arrangements which they had managed to turn to their own advantage (cf. Jones 1979). So machine operators and setter-operators would be at best indifferent, and at worst hostile, to campaigns to abolish payment-by-results.

From the point of view of organised labour in Italy, however, the 1969–70 protest movement, and the related work-enhancement movement, led to one overriding gain. That was the resulting reforms of industrial-relations institutions. National-level union activities gained, for the first time, legal recognition and a corresponding degree of legitimacy in national politics. Moreover, collective bargaining was finally recognised and institutionalised at enterprise level. The chief effects of the legislation relevant to our concerns were the provision of workers' rights to selected company information, and the setting up of representatives, *delegati*, in a new factory council (*consiglia di fabbrica*) system of representation. This system has workgroup constituencies with the right to elect and recall *delegati* to the factory council. It replaced the patchy, more centralised workplace representation of the *commissione interne* system; an institution from 1906 which had survived the decades of Fascist hostility.

The factory councils can function like a collective bargaining forum for all work and employment issues. In strongly unionised plants, the employees' side of this arrangement might be dominated by union representatives. The organisation of NC and related technologies has not, in the main, faced radical challenges from the factory councils. However, from 1970 the representative system gave the *delegati* a bargaining influence over aspects of its usage. From an Anglo-American perspective this influence was used in an eccentric fashion. But the impact that did take place would probably not otherwise have materialised. In the absence of national surveys, case studies are the basis for these inferences. These studies also give us the best indications of specific outcomes and patterns of union involvement.

NC work roles and egalitarian solidarity

That union strength of itself did not automatically mean shopfloor controls was shown by evidence from a sample of engineering firms in the CGIL union stronghold of Bologna. These large and medium-sized firms in that province tended to leave programming as the preserve of technicians rather than operators (Ascari et al. 1984). In general only the most skilled and experienced shopfloor staff retained or improved task autonomy in these firms. The more specialised workers lost opportunities to intervene because technicians exerted much more control. The detailed processes of change in such plants is unclear. Evidence from other individual case studies indicates, however, the limits of union pressure and work reform strategies on the application of NC. Two firms with similar production profiles adopted NC over a five- to ten-year period. Each ended up with similar levels of competence amongst operators, despite having different levels of union involvement in the technological changes (Brivio 1985).

In the largest of this pair, union bargaining won agreement to raise the skills of the operators to incorporate programming tasks. In the event, however, managers argued that the complexity of the programs precluded such a delegation of responsibilities. At first, the programmers even took over the detailed editing of the machines at the 'prove-out' stage. Eventually, however, management were forced – at least tacitly – to cede editing to operators. Pressure for this delegation came from the individual grievances of both experienced and newly recruited operators at the restricted career paths in shopfloor work which accompanied centralisation of tasks. In the smaller firm, the union was not involved in reorganisation for NC. Nevertheless, operators dealt with program editing, in liaison with technicians, from the initial installation.

These two cases reflect the organisational influences in the Anglo-American experience. Small firms adopt operator involvement for practical and organisational reasons linked to their size. Although union influence is relevant in larger firms, radical union schemes for shopfloor programming roles were resisted by managers preferring bureaucratic centralisation. Unionised workers can still, however, claw back some controls when motivated by the other grievances entailed in such centralisation. Thus Italian unionism's newly won strength did not translate spontaneously into control over key NC tasks. The potential influence was there, but the general strategies – influenced by the principles of egalitarian solidarity – were too blunt a weapon to counter managerial concerns about organisational deficiencies. Yet unionised resistance was potent enough to block a monopoly of programming tasks by technical staff.

Union policies in two larger Bologna firms highlight the contradictory dynamics of egalitarian solidarity during this phase. Sasib and GD are large producers of cigarette packing and general packaging machinery, both with strong union organisations. Murray studied Sasib in 1982 (Murray 1984), and GD is cited in Sabel's 1982 overview of NC usage. These firms are not necessarily typical of Italian plant politics, but the intensity of egalitarian unionism there – the union at Sasib is seen as a training school for left-wing political figures – shows the logic of this politics most clearly. Union influence at both plants waxed in the militancy of the early 1970s, but waned later as Sasib tried to cope with the flux and variety of production by extensive sub-contracting, and substantial CNC investment. Throughout these rationalisations of production, the union at Sasib campaigned on several fronts to improve work roles and careers for different classes of worker. By 1985, my interviews showed many operators were, *de facto*, in charge of completing and editing part-programs at the machine. Although – like Brivio's case study – union policies did not directly achieve this. At both firms the unions' main response to the deskilling threat of NC was *horizontal* expansions of skills – more versatility between different types of machines.

In aiming at increasing all-round versatility in the 'horizontal' dimension of machine-setting and operation, these schemes were consistent with the 'mass worker' model. In several respects this logic was inadequate. If managers did seek to centralise programming and detail planning the horizontal expansion of manual skills would not check such intellectual deskilling. Yet, in other respects, the unions' counter-proposals were effective. In Sasib, for example, it was clear that many machine shop workers were not the equivalent of the traditional, all-round, British craftworker – someone trained to operate diverse types of machine-tool. Some were highly skilled, but only on a range of machines within one category. There was a further risk of the newer operators failing to reach even that level of versatility. So the union rejected management proposals to employ workers who were mobile only between machines within certain categories. Instead, collective workgroups possessing responsibility for a variety of different machines were proposed and piloted (Murray 1984, pp. 96–102). Similar, union-inspired, expansions of machinist roles were begun in part of GD's machine shop, after initial management opposition (Murray 1984, pp. 140–2).

Two specific factors explain this simultaneous broadening and limiting of union perspectives for work organisation. The principles of egalitarian solidarity aimed at abolishing a specialised division of labour. These ideas inspired the aim of broadening manual work roles to increase the horizontal span of skills; what work organisation experts call 'job enlargement'. But such a perspective did not arise from any experience of the likely centralisation and hierarchisation of conceptual control with NC technology. It was, therefore, like AUEW policy in Britain: self-limiting in so far as it lacked a significant '*vertical*' extension of tasks and skills. The second reason for the self-limiting character of job enlargement policy was the social and technical relationships between the unionised manual workers and the technicians with NC planning and programming roles.

In many firms, these technical workers are poorly unionised and Italian unions have long sought to increase their membership and involvement. The logic of egalitarian solidarity demanded that they be drawn into policies for reducing task specialisation. Unlike the USA there are no legal obstacles to such a strategy. Neither are there separate occupational unions for 'technical' workers, to organise rival union identities: a factor which, in Britain, it may be recalled, meant competition between technicians and shopfloor workers for programming responsibilities. Yet the Italian metalworkers' unions have largely failed to unionise technicians. The latter have enjoyed superior social status. They have not felt themselves to be 'proletarianised', as some of the unions' Marxist assumptions would predict. Moreover, they have often enjoyed the economic benefits from possession of relatively scarce skills and qualifications (Rampini 1982, p. 14, cited in Murray 1984, p. 104; Rieser 1988, pp. 9–16). Thus, shopfloor unions have been unable to get joint policies, with the technicians involved, to control and organise NC work. Yet, unions' political assumptions have also negated unilateral efforts to take over programming tasks. These assumptions preclude actions that might antagonise the technicians, in favour of winning them over to the cause and to membership (cf. Murray 1984, pp. 81, 104).[9]

As in Britain and the USA, Italian unions have not developed direct policies and campaigns which have reversed the Tayloristic logic of using NC to minimise the planning decisions of shopfloor workers (Rollier 1986, p. 122). In the absence of national data our best estimate must be that they have secured results closer to the British levels of influence than those in the USA. As in Britain, general union strengths have contributed to many workers assuming, or clawing back, some programming. But the conscious campaigns of Italian unionism in the 1970s were focused elsewhere into the strategy of transcending the degradation of work that arose from the earlier, post-war Taylorisation phase. The campaigns' main content was the general elevation and equalisation of skills amongst the manual workforce. For social and political reasons, also linked to the philosophy of egalitarian solidarity, there has been little impact on the boundaries with technicians' jobs.

Thus the effects on jurisdiction over programming have been limited. Nevertheless there is a residual capacity for such policies in the institutionalised powers of the unions. The system of *delegati* and the *consiglia di fabbrica* has made it possible to focus union concerns on the grievances and sentiments of the immediate workgroups, such as the machine shop sections affected by NC. However, some accounts suggest that the attempts at egalitarian conditions of employment led to rigidities in the definition of jobs (Rollier 1986, p. 121). The participative dynamics may also have atrophied since the recessionary 1980s (Heine 1986, p. 193). Nevertheless the significant influence that these institutions have had – as well as the continuing preoccupation with the effects of the inequalities of work roles – should also be recognised as, ultimately, another perverse consequence of the Taylorisation policies of managements in the early post-war period.

Conclusion: numerical control – a Pandora's box?

With the partial exception of the IUE in North America, unions in all three countries lacked highly organised policies and detailed philosophies to claim effective control of NC/CNC operations. Nevertheless, union concerns with allocation of workers to NC jobs, with pay, or with the general principles of occupational specialisation, had an impact upon the deployment of manual skills to NC and CNC. Often these factors operated only indirectly, as in the USA, and there were specific differences in the levels and areas of shopfloor involvement in the programming, editing and overall control of NC and CNC. However, if the only influential factors had been technical and organisational factors, managements would most probably have attempted to use much lower skill levels. Shopfloor customs, union policies and industrial relations institutions all served to maintain skill inputs despite some conditions favourable to deskilling. The marked inter-plant variations, in both the exact mixes of skills utilised – even in the same country – and the corresponding methods, are also partly attributable to specific political dynamics between management and labour within the plants themselves.

So, by the beginning of the 1980s, NC's radical Taylorising potential was being diluted within the social organisations of skill, and weakened by long-standing technical problems of metalworking factories. The precise combinations of skills and responsibilities with the technology varied from country to country. To a lesser extent they could also vary from plant to plant within the same country. The variations did not necessarily correspond to batch sizes and quality requirements. Rather, it was because in some firms there were contests over the exercise and deployment of skills, and because these contests were conducted and resolved in different ways at the plant level; notwithstanding the general conditions deriving from national industrial relations institutions. Despite this 'plant particularism', some general implications of NC as a technological panacea were emerging.

For those willing and able to recognise it, CNC in particular was undermining rather than realising the Taylorist logic. Incentive payment arrangements, detailed prescription of tasks, and centralised decision-making in a hierarchy of competences, all began to seem either irrelevant or counter-productive to the efficiency of computerised machining. Skilled inputs of various levels and kinds were necessary at the machine itself, and not just for programmers and planners in the technical departments. Nor could these skills be prescribed and controlled by the old fixed schedules of scientific management. The craft skills still required were often tacit, rooted in sub-conscious conceptions and varied experiences, and so not easily articulated and scheduled (Jones 1984; Jones and Wood 1984). Task performance was not easily compatible with the vertical channels of communication and authority of Taylorist bureaucracy as skills would be needed in unpredictable and irregular circumstances and required lateral and *ad hoc* forms of interaction between machining, setting, maintenance and programming personnel. Depending on the character of the local and national-institutional factors, the transition from NC to CNC once more revived the logic of workshop, rather than factory organisation.

Some methods of using CNC were pushing Taylorism to – or even beyond – its conceptual limits; but no necessary and general shift to decentralised, worker-centred organisational paradigms followed. Many managers were equally capable of drawing two different conclusions. One inference was the pragmatic acceptance of deviations, such as operator editing of programs. Such operator involvement was a necessary evil to be tolerated while remaining operations continued along conventional Tayloristic lines. The other, more radical, judgement was that NC/CNC provided insufficient centralisation of control and automation of machining operations. By the early 1970s automation-oriented technologists were developing and advocating higher levels of computerisation. Such systems would control not just individual machine-tools but entire sections, as in more advanced versions of integrated cells, and plants; linking these to other functions such as design and production planning. This cybernation promised the Holy Grail of shopfloor solutions: bringing the flow production principles of mass production to the batch production factory. Fordist concepts, hitherto subordinate to an overall Taylorist organisational paradigm in batch metalworking, began to strengthen their appeal to managers.

III Cybernation and flexibility

Overview

Futurist assumptions dominate much academic discussion of the implications of the move to computer integrated production. Such influence extends to radical critics, such as the Marxist historian David Noble, for whom the promotion and propagation of CIM is the inevitable consequence of long-standing, grand designs, stretching back to the development of NC, to extend corporate control at the expense of worker resistance. Futurist thinking has swayed the managerial literature even more. Accepting the future path of cybernation as already mapped out by technologists, the task of managers and their advisers is seen as devising and promoting appropriate organisational and human relational infrastructures to realise the technology's full economic potential. Analysis is consequently predicated on the assumption of a huge, beneficial potential locked up in the technology; which the right business and management policy recipes can realise.

More active participants in these and related developments – usually engineers – often have a hard-nosed Fordist, or Taylorist view of their potential.[1] Yet management academics often view such approaches mainly as contrasts with the essence and potential of FMS – as innovative, post-Fordist versatility with skill-based flexible work roles. Bessant, for example, emphasises the tendency to use FMS for 'operational flexibility': gains of task-minimising, time-saving, machine use-maximising etc. Or 'even, in many cases, for traditional cost-saving motives' (Bessant 1991, p. 110). Yet, Bessant then goes on to measure such cases against more innovative uses of FMS, 'optimal arrangements', 'economies of scope' coupled with less specialised and more autonomous workgroups (Bessant 1991, pp. 116–26). An unfortunate effect of such mixing of analysis and advocacy is to obscure the reasons for the allegedly sub-optimal, but prevalent, types of FMS use. More importantly they amount to a crypto-Taylorism: a presumed 'one-best-way' to organise production.

The following chapters aim to avoid this logic by examining the shaping of FMS usage in the USA, Japan, Britain and Italy by national industrial structures, socio-political histories, and institutions in the industrial relations sphere. In this way we can understand the reasons why firms use FMS as they do, without judging them as British and US deviations from, or (wrongly) Japanese approximations to, a notionally superior essence. This is not to deny a potential for more innovative and versatile uses in FMS. After describing, in the next chapter, the persistence of Fordist production criteria and Taylorist work organisation in the USA, chapter 7 examines in detail a deviant North American FMS case, corresponding to Piore and Sabel's model of flexible specialisation. The value of this separate analysis is that it shows the variations in concrete circumstances which may induce 'economies of scope' or flexible specialisation. The restrictive grid of a notionally superior alternative also obscures each national context.

6 The cybernated factory and the American dream

Introduction

During the 1980s I criss-crossed North America by plane and car visiting Flexible Manufacturing Systems (FMSs) in aerospace, agricultural, construction and other machinery-making firms. I observed at first hand nine of these systems which bring the operations of automatic machine-tools, the automatic transport of workpieces and tool-cutters, and their scheduling, under central computer control. FMSs can justifiably be regarded as colonies of cybernation, or – to use the technologists' jargon – of Computer Integrated Manufacturing (CIM) within larger conventional plants. It is justifiable because FMSs consist of these three separate automatic systems: machining operations; tool and workpiece conveyance; and overall scheduling. FMS therefore represents the climax of the piecemeal automation that began with NC and a projected 'final solution' to the idiosyncratic complexities of batch production. CIM and FMS aspired to achieve factory-like flow methods of control and production within the constraints of small-batch product variety; a Fordism for the 'metal-bashers'.

In these plants I conducted semi-structured interviews with managers, engineers, workers and trade unionists (see Jones 1985b, 1989; Jones and Scott 1987). The resulting data partially corroborate the arguments of other writers (Noble 1984; Jaikumar 1986) that the claimed, or potential, flexibility and efficiency of these forms of CIM are not realised by the US firms using them. The technology was combined with an over-specialised pattern of work organisation, made up of restrictive occupational responsibilities, that seem ill-suited to the needs of controlling, adjusting and maintaining complex interacting systems. There is a consistency in the reasons for sub-optimal use of the systems between these case studies and other research on US FMSs (Jaikumar 1986; Blumberg and Gerwin 1981). Yet US firms' persistence with these arrangements and methods raises the more important question of why the managers concerned seem largely satisfied with the restricted and unreliable operation of their FMS?

In view of its limited popularity and limited efficiency, why was this particular technology developed and marketed in the first place? The simplest answer is that in North America in the late 1970s and early 1980s the main inefficiencies in batch production factories were believed to be operating costs and the weakness of direct managerial control. These deficiencies were increasingly associated with gaps and delays in the productive process, for which computerised information and control systems, or CIM, began to be promoted as the solution. Yet even small-scale CIM applications, such as FMS, have proved to be costly, complex and difficult to marry with other economic and organisational practices.

As with earlier attempts to reform batch production, the original aims also proved over-ambitious. CIM systems are still being adopted. However, relatively well-tested CIM applications, such as FMS, have not generally – in the USA – achieved all the gains originally intended; and the efficiency of British installations has been questionable (Jaikumar 1984; Dempsey 1983). Predictions that 50 per cent of new machine-tools will be for use in computer controlled systems (Olling 1978) have proved to be far too optimistic. Moreover, the types of efficiency originally targeted for FMS technology may have become relatively less important in the changing economic and business conditions of the late 1980s and 1990s. Where these call for market responsiveness and product innovation large-scale technologically intensive FMSs in the USA look environmentally ill-adapted.

On these kinds of criteria they may be out-performed by less technologically ambitious uses of FMS in Japan. They have also been rivalled by a resurgence of workshop-type operations, assisted by more piecemeal use of computerised production, such as CNC machines, often in localised networks of smaller and medium-sized firms in Europe. These widely differing uses of computerisation in small-batch production, have important implications for ambitious US users' strategy of cybernating operations, and for current debates about the Fordist, post-Fordist or Taylorist character of new manufacturing paradigms. The precise and prospective impact of differing modes of computerisation has been indicated in the previous chapters on NC and CNC. To complete this analysis also requires detailed comparisons between the use of FMS in the USA, Japan and Europe. This chapter will describe the patterns of implementation and organisation of FMSs in the USA. It seeks to show that a combination of Tayloristic administrative paradigms, the appeal of Fordist production management, and the continuance of contractual individualism in industrial relations is constraining underlying technical flexibilities within Fordist limits. But first we need a historical sketch of the evolution and logic of the FMS form of cybernation.

Technological trajectory: pursuit of the workerless factory

What were the principal aims in the propagation of CIM technologies such as FMS? Clarification of these 'push factors' is important for they bear on both the utility and adoption of the systems and, more precisely, on the contrast between designers' and advocates' picture of computer integrated control – the one purveyed in the business and technical media – and actual operations. The two most influential and contrasting explanations of the ostensible purpose of such systems can be termed the 'economies of scope' and the 'labour control' theses. The former sees the aims of computerisation as maintaining, and even extending, the traditional product range of batch production via greater operational efficiency and reduced indirect costs. From the labour control perspective, on the other hand, the aim of technological renewal is the direct managerial control of operating procedures by the virtual elimination of production workers' discretion and decisions, with labour cost reductions as the financial justification.

David Noble's bravura history of the development of computerised metalworking in the USA, has argued forcefully that an alliance of the US military establishment and large defence contractors promoted the latest CIM phase of factory automation in order to transfer control of production operations from unreliable skilled machinists to managements. Nobel describes the evolution of an ideological crusade – what I have termed 'futurism' – for total automation amongst US technologists and industrial managers since the Second World War. He links this common belief system from speculative 1940s' prophecies – for the spread of automatic controls used in the materials processing industries – to the US Air Force's (USAF) championing of the first innovations in numerical control in the 1950s; and to the same agency's campaign for CIM, through project ICAM and its derivatives.

ICAM – Integrated Computer Aided Manufacturing – aimed to develop both the software and the operating expertise for a hierarchical control structure linking the design (CAD), material and equipment planning (CAM) and NC machining sub-systems of batch metalworking. ICAM, says Noble, sought technological solutions to solve a variety of worker-related problems: to cure the problem of out-of-control labour forces; to 'try to reduce the enormous indirect costs that have resulted from the effort to try to reduce labour costs and remove power and judgement from the shop floor' (Noble 1984, p. 332); to realise the dream of total control of manufacturing (p. 332); to stop workers' involvement in adapting and correcting computerised machining; and to convert human 'know-how' of operations into computer-usable forms of information (p. 333). An obsession with higher levels of automatic, rather than worker control, over methods and output, resulted in CIM projects for the realisation of ideological fantasies with no strict economic advantage. Extra overhead, fixed capital and running-in costs outweigh any savings on labour costs (Noble 1984, p. 343). Moreover, in the USA, firms only invest in CIM because USAF and other military agencies' contracts effectively subsidise their losses in setting up working demonstrations of such technology (Noble 1984, p. 340).

Noble's account usefully exposes the ambiguities and inconsistencies in the purposes of CIM technology. It is, however, of limited value as a general explanation because, firstly, its emphasis on the major role of the US military-industrial complex makes it relevant only to that country. Yet there are promoters, makers and users of CIM technology in most advanced industrial societies; notably in some – such as Japan and West Germany – with negligible military demand. Secondly, it overlooks the fact that US machine-tool firms and their customers adopted and developed FMS both before, and independently of, the ICAM programme. So unless we accept that the ideology of automatic workerless factories has mesmerised all US managements, there must be other influential motives. Noble's thesis is plausible because ICAM, and the military's push for FMS and other CIM forms, coincided with a revival of interest in their applicability; after a fallow period of about ten years from when industry tried out the first systems. What distinguishes these first and second phases of interest is a sharper specification of the properties and relevance of the technology in the later period. In other words, it was not until a different production paradigm developed, amongst suppliers and user firms' managers, that FMS, at least, was 'seen' as a feasible technology.

As chapter 3 explained, the first operating system approximating FMS principles was not American at all, but the British Molins firm's 'System-24'. Under the enthusiastic influence of their Technical Director, D. T. N. Williamson, Molins had tried in the mid-1960s to market System-24, and some of its machining processes, as an adjunct to the firm's main product line of cigarette machines. Although the system was marketed in the USA (*American Machinist*, 11 September 1967) there were no buyers. Around the same time Cincinatti Milacron, a major US machine-tool manufacturer also developed a computer-linked machining system 'but never wrote a single order' (*Business Week*, 3 August 1981, p. 60). The Sundstrand Corporation supplied two versions of a machining centre, to Ingersoll Rand and to Caterpillar in 1970 and 1971; but no other external orders materialised.

Thereafter there was a lull until the later 1970s when systems with more developed computing controls began to be bought. Contrary to Noble's account, this did not take place only amongst sub-contractors to the military. The main buyers were mainstream users of machine-tools in the agricultural, transportation and construction equipment sectors; from vendors such as the Kearney and Trecker (now Cross and Trecker) Corporation, whose systems were the first to be labelled FMS.

Commentators and participants in the 'phoney war' phase of FMS identify a range of technical and commercial reasons for non-adoption: inadequate software; costs of link-up to central computers; the need for costly and time-consuming software modifications; imprecision in quantifying financial savings in investment proposals (*Business Week*, 3 August 1981). These points were certainly relevant considerations for potential investors, especially the greater complexity of reprogramming pre-microprocessor computers. However, they are not a complete explanation of non-adoption; for the same drawbacks – especially reprogramming costs and the vagueness of quantification for investment justification – are still voiced today. A more influential factor was probably that promoters of the technology – in the 1970s the machine-tool manufacturers and not the military – lacked the concepts and language to articulate its relevance to corporate and manufacturing management. During the initial phase of computerised systems this paradigm weakness can be summed up in the virtual absence of one word: flexibility.

In 1983 the US Patent Office awarded retrospective patent rights to Molins, and their lawyers, as originators of System 24: 'what has come to be called an FMS nowadays'.[1] This award, which has since been contested and reversed in the courts, overruled competing claims by Cincinatti Milacron and White Sundstrand. Yet, neither Molins nor other early systems spoke of Flexible Manufacturing. Williamson's 1967 description of System-24 for *American Machinist* only managed: 'a conveyor-linked line of specialised, contouring NC machines with automatic toolchanging – controlled by an on-line computer' (Williamson 1968, p. 145). Likewise, White Sundstrand's set up at Ingersoll was simply a 'Heavy Machining Centre'. Cincinatti Milacron's 'Variable Mission Manufacturing System' title came closer to the flexibility concept. However, an MIT research team – set up in 1973 to find a computerised synthesis of job shop and mass production techniques for small-batch machining – chose the generic, but less marketable, term Computer Managed Parts Manufacture (General Accounting Office n.d.). The GAO's comprehensive report to the US Congress, circa 1975, discussed neither Flexible Manufacturing Systems, nor flexible capabilities. Its 150 pages refer just once to 'computer-controlled flexible manufacturing systems' (General Accounting Office n.d., p. 31).

These early systems did not emphasise the role of adaptable controls partly because reprogramming was complicated; but also because they were initially conceived as a means to bring Fordist economies to batch production (Cook 1975), rather than as means of refining the versatility needed for making shorter runs. Promoters as well as users remained within a Fordist paradigm. Thus in 1979, George Hutchinson, an academic authority on FMS, and one of the first to refer to 'Flexible Manufacturing Systems', wrote that 'variance is the enemy of efficiency in Automated Batch Manufacturing Systems' (his term for FMS and similar systems: Hutchinson 1979). A review of the computerised systems in operation in the mid-1970s claimed: 'Job lot manufacturers in the US need production equipment that is as cost effective as the highly automated transfer lines employed by high volume producers' (Olling 1978). The same text ranked an undefined quality of 'flexibility' at the bottom of a list of seven cost-saving benefits.

Contrary to Noble's account the detailed objectives of the ICAM mission, although dominated by the concept of integration of plant functions, was composed of different, and potentially inconsistent aims. Labour displacement and direct management control was only one facet. In addition to changes in manufacturing strategy, the 'direct benefits' of ICAM technology included both 'more efficient production management' and 'reduction in the time between design and delivery', as well as both 'reduction of maintenance costs and parts inventories' and 'increased production rate and *flexibility*' (Mayfield 1979; emphasis added). For, by the time that the ICAM prospectus was being written, the direct cost-cutting assumptions of the productivity paradigm were being overtaken by ideas about 'economies of scope'. As a result, from its inception, ICAM, as well as other commercial proposals, was promoting divergent, partly incompatible, and possibly misleading, manufacturing objectives: cost-minimising efficiency and operational versatility. As we shall see in the following review of case study evidence the operation of many of the FMSs initiated in the early 1980s reflected these contradictory tensions.

The economies of scope perspective laid primary emphasis on the capability to extend the *range* of items produced within given or reduced cost parameters. It departed from mass production principles in:

> rejecting the notion that greater production volumes display lower unit costs than do lesser volumes . . . the new technical capabilities rest on economies of scope – that is efficiencies wrought by variety, not volume. (Goldhar and Jellinek 1983, p. 142)

In the new paradigm the reduction of direct costs, such as labour, virtually disappears. Instead, it promises more qualitative benefits such as:

> *Extreme flexibility* in product design and product mix . . . *Rapid response* to changes in market demand, product design and mix, output rate and equipment scheduling . . . *Greater control*, accuracy, and repeatability of processes, all of which lead to better quality . . . and reliability . . . *Reduced waste*, lower training and changeover costs . . . *Greater predictability* . . . which make possible more intensive management and control . . . *Faster throughput*. (Goldhar and Jellinek 1983, p. 142; original emphasis)

In a similar vein Piore and Sabel emphasise the potential of computerisation – provided other institutional conditions are met – to realise Flexible Specialisation as a business strategy; in which retooling and resetting are simplified to make small-batch production as economic as mass production. They emphasise a similarity, between historical and overseas cases of Flexible Specialisation and the business analysts' identification of: 'shorter product life cycles, more productive proliferation, smaller order quantities. . .' with the technological capability '. . . to react, to be flexible' (Piore and Sabel 1984, p. 262). As we shall now see, American firms adopting FMS largely failed to apply FMSs for flexible specialisation and economies of scope because of technological, operational and institutional conditions.

The *institutional* reasons were, as with the utilisation of stand-alone NC tools, mainly the organisation of managerial control and the contractual individualism of the industrial relations system. The *operational* reasons were the attempt to gain Fordist direct-cost reduction goals and at the same time gain flexibility of production ranges. Piore and Sabel describe computerised manufacturing as a kind of magic mirror providing images of unpractised flexible production. A manager from one of the machinery manufacturers interviewed for this book used the metaphor in a more sardonic way. He characterised FMS buyers as managers 'who think that FMS is a magic system, that it will do everything'. This magical mystique exemplifies the ambiguous *technological* properties that have problematised actual operations. The promotion of CIM and FMS in the late 1970s and early 1980s stressed enhanced variety of product range and rapidity in introducing new products. But these claims underrated the enormous complexity of writing basic software and of operational programming and reprogramming necessary to achieve such aims (Zygmont 1986).

Nathaniel Cook, of MIT's Computer Managed Parts Manufacture team believes that development engineers had been trying to get a quantitative measure of flexibility, 'which was what was needed', but they had become completely bogged down at that point. Plausible performance levels were gained with set part-families that were 'stationary in time' but, with broader selections, 'the thing degrades'.[2] Flexibility in the range of products that can be simultaneously machined, and the systems' capability to accept and process new items, requires the calculation and co-ordination of a huge range of different variables. By the mid-1980s, the difficulties of developing software to encompass this range, together with the lack of tool-cutting reliability to achieve the other goal of minimal labour, were undermining experts' confidence in the extent of CIM-based flexibility: 'The whole thing is floundering. Ten years ago we thought that these systems would just go "oomph". Not now.'[3]

Using FMS: the magic machine vs. the sorcerer's apprentices

The FMSs that were installed in increasing numbers in the early 1980s had to cope with the practical consequences of the contradiction between scope economies and labour control; between flexibility of product range and minimisation of labour. Survey evidence indicates that most managements were already biased towards minimisation of labour and other costs. Engineers' production paradigms remained predominantly Fordist. A comparison between similar FMSs in Japan and the USA showed that each US installation made an average of 10 parts, while the Japanese counterparts made an average of 93. Where the US volume for each type of part was 1,727 the equivalent Japanese figure was 256. While the US firms introduced an average of one new part into the systems' capabilities each year, the Japanese introduced 22. These figures from Jaikumar's study (Jaikumar 1986, p. 70) may not be representative of all Japanese FMSs, as he looked only at machine-tool firms. These may be expected to have more in-house expertise in automatic machinery to devote to FMSs, and more interest in developing their capabilities.

More significantly, US firms' use of a FMS – for 'high volume, standardised production' – is associated with restricted roles of the remaining production staff, especially their exclusion from involvement in programming during both installation and any subsequent software modifications. Working knowledge of the systems is specialised, compartmentalised and regulated by procedures. In particular, other studies (Graham and Rosenthal 1985; Roitman et al. 1986) confirmed that even where FMS managers strove to combine flexible capability with cost reduction goals, the division of tasks, skills and responsibilities between various managerial and production staff functions blunted their efforts. In the mid-1980s, I interviewed at nine FMS plants in various firms and regions across the USA (for details see Appendix). These were a representative spread in terms of locations, ages and sectors, from the broader pool of about twenty FMSs then operating in the USA. Supplementary interviews were also conducted with academic engineers, industry experts and key personnel in firms selling FMS. From these enquiries three firm conclusions stand out.

Firstly, the systems were not purchased to achieve one single overriding goal. The most commonly *mentioned* aims were: simplification of operating procedures, direct labour inputs and inventory; as well as the related cost reductions. However, each firm referred to a variety of motives for FMS investment. Although these quantitative aims were dominant they were also often accompanied by references to qualitative flexibility, such as the adaptability of the system to other uses. The *second* conclusion is that the majority of the systems were actually operated in a much more limited way than was suggested by either the firms' aims, or the capability of the FMS. *Thirdly*, the work tasks were organised in a hierarchical manner which reflected both managements' labour and production control philosophies and the continuing culture and practice of contractual individualism in industrial relations. In case study research small sample sizes mean that distributions of the numbers exhibiting one characteristic rather than another are not very meaningful. The following explanation of the aims, operations and work organisation of the FMS plants therefore emphasises contrasts between different *types* of firm in each of these three areas.

Strategic aims

It has to be recognised that the most rigorous of interview and survey analyses of investment decisions are only constructing a *post festum* ranking of a highly complex process of motivational and micro-political conflict (Jones 1989b, p. 473). Individual firms' and managers' explanations may well be after-the-event rationalisations of what were originally imprecise and eclectic aims, motives and expectations amongst different managerial specialisms. In one case it is clear that purchase of FMSs later became rationalised through, and subsumed beneath, a utopian vision for computer-integrated rationalisation of the entire corporate manufacturing capacity.[4] In five of the nine plants in which I interviewed, the most-cited reason for FMS purchase was reduction of operating costs, principally work-in-progress and labour; a finding matched in a single, more detailed case study by Thomas (1994, p. 51). FMSs were also often said to be selected instead of other types of plant, such as individual NC machines or fixed transfer lines, as adoption of these would mean lower equipment utilisation, because of the need to produce medium-sized batches.

However, four firms referred to some form of 'economies of scope' flexibility. Management in these plants had no common vision of what flexible manufacturing should consist in. Rather, they had a general awareness of the technological versatility of FMS and identified specific attributes thought to be useful to their business. In this respect the capability to adapt more quickly to future model and design changes was the most common theme, together with, in one case, the ability of the system to process a wide variety of parts – they claimed over 200 – on a near-random basis. In general, and in practice, these claimed advantages were mostly unproven, or difficult to corroborate. The exception was the 'Alpha' plant, the oldest FMS in the sample, which the next chapter describes in detail.

Thus the general evidence from these cases partly confirms the 'continuist' labour control thesis of Nobel and other writers of that tradition. It is also partly consistent with Jaikumar's findings. As with the latter's broader survey the reduction of operating costs was the predominant rationale; and, as claimed by Noble, minimisation of labour inputs was a major aim. Yet it was as a *cost* that management saw labour as a factor to be reduced; and overall it was no more central than the other main cost reduction aims: work-in-progress and unused machine time. Contrary to Jaikumar's diagnosis (Jaikumar 1986, p. 71), and the implicit reasoning of Noble, this was not due to Taylorist managerial rationality. It is much more consistent with the different Fordist approach of minimising breaks and variations in the flow of production by automation and tighter pre-specification of methods and product features. The important issue is *why* awareness of the systems' versatility – range flexibility – was rarely converted into practice? Jaikumar and other analysts only address this point conjecturally. Also unresolved is the related question raised at the beginning of this chapter – why plant managers seem so satisfied with sub-optimality in their FMSs? The operating methods and approaches pursued partly explain these weaknesses in US firms.

Operating practices

Consistent with Jaikumar's assessment the range of product parts made on the FMSs was rarely large and varied. There were considerable differences in the numbers of types of part and the degree of similarity between the different part 'families', into which the individual parts were grouped. The FMSs making larger parts, for railroad and construction equipment – 'Locos', 'Tractors', 'Diggers' – would process as few as ten types of part. Yet the two aerospace plants studied – 'Bomber' and 'Aerosystems' – claimed a capacity for random assignment of over 500 part types in sixteen different part 'families'. Several FMSs had been bought specifically to handle the machining requirements for a new product. In that sense they were to a certain extent 'dedicated' in true Fordist fashion. However, as the deviant case of the FMS at Alpha – one of the agricultural equipment firms – will show, this need not necessarily preclude subsequent adaptation.

More consistent with the bias towards limited product range was the policy of intensive utilisation of the systems. Several of the firms admitted that the period of running in, and initial preparation, of the FMS was longer than anticipated. However, they subsequently gave priority to keeping the system running, or maximising 'up-time'. Partly, perhaps, because of these initial delays – and the unexpected down-time – anything that might mean a diminution of machining time was resisted. The crews of FMS workers were given tight job descriptions which inhibited, and often prohibited, reprogramming of the machining and scheduling software. The FMSs were often dependent upon external technicians and engineers for programming – and sometimes also the general maintenance department of the plant – and the technician in charge of the computer control of the entire system was normally preoccupied with problem-solving in order to keep to the existing machining and scheduling routines. Even if management had wanted to introduce many new part designs into the system they would have had to accept a multiplication of these 'normal' operating difficulties.

The latter included computer failures due to the electricity generated by mid-summer storms, the failure of the scheduling software to keep up with *ad hoc* rerouteings and reworkings of problematic parts, as well as more mundane tool breakages. Minute undetectable variations in the dimensions of parts due to their automatic clamping on pallets and workpieces caused particular problems. This last point illustrates a general conclusion about the capacity of FMSs to generate dynamic error processes (Gerwin and Leung 1980). Slight departures from standard tolerances and dimensions, which might be insignificant or passable when occurring on one machine, are compounded in FMSs as the faulty component is passed from one machine to another accumulating further deviations at each operation.

So both installation and day-to-day problems combined to make continuous utilisation of the equipment a priority. In addition, as Jaikumar also suggests, several companies' tight financial controls pressured FMS managers to prove the productivity of the systems in terms of maximising up-time. This combination of factors exacerbated managers' predisposition to discourage any activity which might reduce continuous operation of the system: including the experimentation and development of new programs necessary to extend the product range, or render the system more amenable to some future input of new designs. Contrary to Noble's thesis, system inefficiencies were not tolerated in a cost-indifferent manner. Rather, it was cost pressures which exacerbated the rigid use of the systems and the avoidance of flexible operation. In this mental and procedural climate the monster metaphor replaced that of the magic machine. In the words of a manager at Aerosystems, a relatively sophisticated FMS, the perpetual preoccupation was 'to feed the beast'.

Work organisation and industrial relations

Only two of the case study firms were not unionised. Interviews with unions provided a check on the information given by managers, and where possible the technical and production workers, and a source of additional data. These supplementary interviews were normally carried out at the, usually adjacent, offices of the union local. The relevant plant or shopfloor union representative, and sometimes other FMS workers, were questioned and also invited to give their views on various aspects of the automation of the plant. These union interviews often provided contrasting information on the history and efficiency of the FMSs. They also, in general, confirmed what the management interviews had suggested: that the US system of collective bargaining, and the associated job classification procedures, meshed with the Taylorist administrative arrangements, and added further to the inflexibility and Fordist ethos of the FMSs.

We saw in the preceding chapter how the US system of industrial relations – characterised as one of contractual individualism – gave quasi-legal backing to management policies of excluding unionised machinists from involvement in 'management sphere' programming functions. However, the same system also operates to restrict managements' power to deskill – by replacing skilled and qualified workers with lower-skilled employees – through the workings of the industrial relations practice of 'bumping'. The prevailing system of internal promotion and regulation of lay-offs hinges on the seniority principle: established in the 1930s and 1940s during the period of institutionalisation of industrial unionism in the USA (Jacoby 1985). Consequently in lay-offs, or movement of workers out of rationalised operations, such as automated processes, younger or more recently employed workers are more likely to be displaced. The older, or more 'senior', employee is more likely to be retained, as these workers 'bump' less senior employees out of the affected jobs.

These arrangements act as an indirect brake on the deskilling that might occur through NC computerisation of individual machines, because skills tend to be acquired in serial fashion, through the promotional side of the internal labour market seniority principle. Thus the more senior employees are more likely to be retained; and these tend to be the more skilled.[5] These kinds of movement through the occupational hierarchy may involve changes and negotiations of enormous complexity. In US metalworking plants the contracts governing union–management agreements cover schedules listing the pay, conditions and task descriptions of hundreds of occupational grades. Unions in the USA have exploited Taylorism's principle of finely detailing the task range and responsibilities of each job. These have become the basis of collective bargaining in terms of contractual rights which tightly define both the obligations and rewards of individual workers.

One of the main suppliers, Kearney and Trecker, recommends staffing Flexible Manufacturing Systems with a team of all-round operators and a system manager (Kearney and Trecker n.d.). In practice, most FMSs are organised with highly specialised job descriptions that are similar to conventional machining processes. A typical work crew in the FMS case studies consisted of:

- one or two 'loaders';
- two or more 'operators';
- a maintenance worker or 'repairman';
- a 'system supervisor' or programmer.

The *loaders* were the lowest grade of workers in terms of skills, pay and job security. Their job was to fix the workpieces on to the pallets before entry into the system, and to unload the finished components from pallets. Unless they had 'bumped' into loading jobs from higher grades the loaders were unlikely to have machining or other skills. Sometimes they would be recruited externally and would look upon the loader grade as a 'port of entry', to use a term of Clark Kerr, to higher positions in the workplace. The *operators* undertook the residual tasks of machine adjustment and monitoring that could not be carried out by computer controls. The operators inserted and adjusted the cutting tools in some firms which had no automated tool supply; either because of the extra cost, or because the large number of tools required exceeded the capacity of an automatic tool storage and handling system.

Operators would also carry out routine checks on the dimensions of machined components, although in some of the more recent systems this work could be reduced by the provision of automatic, in-process, gauges. Unscheduled tool wear or breakages would also be corrected by the operators, as well as any physical rerouteing of parts that had to be reworked, or reallocated, because of interruptions to the automatic scheduling and transfer processes. The most typical background for operators was conventional or NC/CNC machinists, normally at a grade near the top of the machinist pay structure. In some of the FMSs the operators undertook more remedial maintenance; and, much more rarely, made minor NC part programming modifications, such as to lengths, or feeds and speeds, of the cutting tools.

The *maintenance* or 'handyman' worker carried out minor *in situ* maintenance or repairs to the mechanical equipment, mainly the machine-tools. Graded at the same or slightly higher level to the operators, this worker's tasks often overlapped with theirs. Some FMSs had no specified, or dedicated, maintenance worker on the FMS. Then these tasks would be divided between the operators and the general maintenance section.

The *system supervisor* or programmer was a qualitatively different position to the rest of the FMS crew. Normally, only he – there were no women in these jobs – would be allowed to change the scheduling programs that controlled the allocation and transfer of work between the different stations in the FMS. Other tasks consisted of carrying out checks on the flow of parts, reading data on the accuracy of machining and tool wear, making *ad hoc* decisions about the rerouteing of parts that needed reworking, introducing different workpieces into the system; as well as liaising with the rest of the FMS crew and the main planning and NC programming sections. Several firms gave the system supervisor managerial authority over the crew; thus accentuating his membership of the managerial, rather than the manual staff. In others, including one which abandoned the dual role, this was regarded as over-complicating the system supervisor's role; then line authority over the crew was concentrated in a supervisor proper, who was sometimes responsible for other sections as well as the FMS.

In general, these job specialisations were rigidly observed. Sometimes, and more regularly in the firms described at the end of this chapter, tasks were shared between the different grades. Yet this sharing took place more often between the core occupations of operators and maintenance workers than at the 'upper' and 'lower' ends of the task hierarchy. Loaders' tasks were rarely taken on by the other workers; but the latter were excluded from any of the controlling activities of the system supervisors. Segregation was for both practical and social reasons. The loaders ranked lowest in the job classifications and, in the plants regulated by tight labour contracts, any of their work done by the operators and maintenance grades would have clearly breached contractual rules. Because of their higher skills and capabilities the operators also disparaged loading as beneath their status. On the other hand, management decrees denied the manual workers system control tasks. As with the separation between NC programming and operating described in the previous chapter, US managements almost always ruled that system programming was an exclusively management task which would violate the boundary between union-regulated, and management-regulated duties if delegated to unionised manual workers.

However, these job demarcations could by no means be attributed solely to union or collective bargaining constraints. There were practical limitations too. In the Gearbox plant, it was argued that when operators had spent time away from the FMS equipment, taking experimental spells in loading or analysing system data, they had quickly forgotten the essential detailed knowledge of the idiosyncrasies of particular machine-tools. Another FMS, in the 'Machinery' plant, was part of a 'runaway' factory set up on a greenfield site without a union presence. This is a significant case because although there was no union, the management still operated with the normal hierarchy of FMS jobs, and refused to cede any programming involvement to the FMS crew. It seems, therefore, that a preference for specialised job classifications was not confined to unions. Management, too, could prefer this arrangement because their managerial practices were embedded in deeply held assumptions about Tayloristic hierarchical specialisation.

In a critical report on FMS organisation two managers from A. T. Kearney (the systems division of FMS supplier Cross and Trecker) have attributed much of the down-time and quality problem to the unrealistic staffing assumptions of FMS users – a claim mirrored in Thomas's case study (Thomas 1994, pp. 62-4). They blame these faults upon buyers, who associated quality and inefficiency problems with labour and assume that as labour was eliminated so would the problems. On the contrary, argue Thompson and Scalpone, the role of the 'human factor' becomes more important *in direct proportion to the numbers reduced.* The machine shop was labour-intensive, but the FMS is information-intensive; and possession and use of information and know-how becomes the lever to raise the utilisation and quality levels of the system. These factors require workers to be well trained, flexible and responsible. Scientific management methods of specialising workers' jobs are counter-productive. Instead, versatility and teamwork across all functions are essential (Thompson and Scalpone n.d.).

These strictures are apt, but misleading about the communication of information entailed by the social structure of the FMS. Especially in my interviews with FMS workers, multiple, but often unacknowledged, contributions by the crews to the running of the systems were reported. It was also clear that informal swapping of tasks and other forms of unofficial co-operation were essential to keep the systems running to schedule. In some cases, the workers had prevented inordinate output and quality problems by working beyond their specified task ranges (for more detail see Jones 1990). At Locos, management even retreated from a specifically negotiated teamworking arrangement after calculating that the extra compensation for this scheme was excessive. The workers complained bitterly of a broken trust in which they had worked co-operatively and conscientiously to help get the FMS established, suspending union work rules, only to have management rescind the agreement after the main problems had been eliminated and the FMS was in full operation.

Perhaps the most important and emblematic feature of these accounts for the neglect of the communication of information is the separation between the computer control and FMS operation functions. Not only are the jobs segregated by administrative and managerial rules, but they are physically separated as well. The FMSs' central computers download the

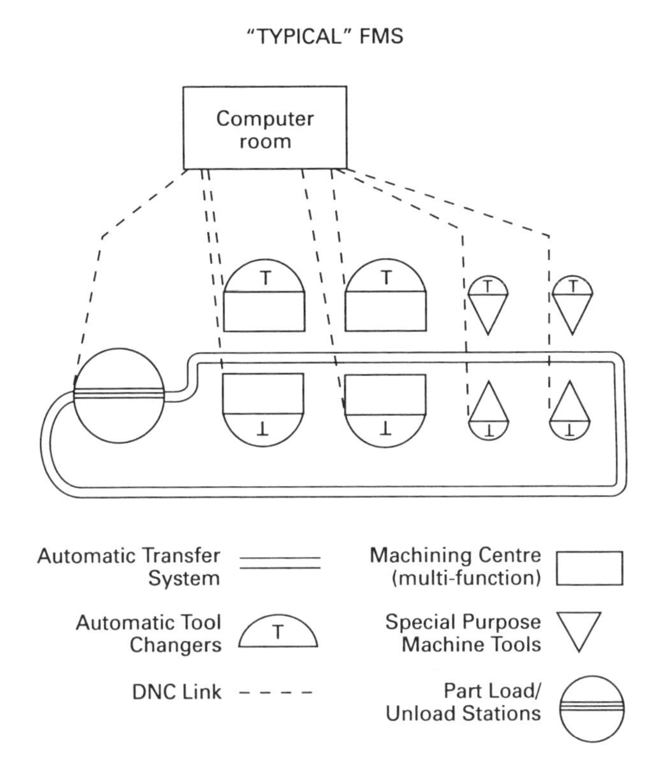

Figure 6.1 Layout of typical Flexible Manufacturing System

part-programs to the individual machines, to provide monitoring and operating data, and calculate and control the schedules of work around the system (see Figure 6.1). In North American FMSs an architectural tradition has located these central computers in cabins resembling air control towers above, and back from, the actual machining area. From these, sometimes air-conditioned, watch towers white-collared programmers and system managers are supposed to survey visually and electronically command the production process and its support requirements.

However, this clinical image jars with other signs of the control process. In recurrent observations of these arrangements it was noticeable how much mundane and human inter-communication was necessary. If they were not initially provided, phones and walkie-talkie radios had to be installed. These were in frequent use, with the programming technician yelling instructions to the FMS crew. Hand-written notices were ham-fistedly taped on to and beside computer terminals, phones and machine-tools. In one FMS which had not, yet, been provided with phone lines to the shopfloor, the sophistication of the information technology was mocked as the system supervisor resorted to primitive hand signals to the watching operators on the floor below. When, as seemed obvious, I questioned the managements of these systems, about the possibility of delegating computer modification tasks to the manual staff to overcome these communication detours, the stock excuses were industrial relations conventions on management–worker demarcations, and a lack of trust in the conscientiousness of the workforce.

Thompson and Scalpone's assessment uses the image of the enchanted broom from Walt Disney's cartoon film *The Sorcerer's Apprentice*. This is used to symbolise the progressive accumulation of faults when a wrongly programmed system, or incorrectly primed workpiece, carries on automatically through the various stages of the system. It is almost as if US managers have watched this scene in their training videos, but interpret it as showing the operating staff as potential sorcerer's apprentices. These cannot be given any licence to experiment with the magic machine for fear that their untutored tinkerings will result in catastrophes on a scale with the animated broomstick.

Conclusion: end of an impossible dream?

The prospects for a US-style automated solution to batch production inefficiencies have receded considerably since the early 1980s. Worldwide aspirations for the extension of CIM and the proliferation of FMSs have been scaled down. The further integration of all the different systems for design, planning, production, inventory and distribution depended on industry acceptance of the all-important communications software based on common standards. Governments and large firms on both sides of the Atlantic have backed the General Motors' solution of MAP (Manufacturing Automation Protocol). However, most potential users have not shown interest in purchasing these interface linkages (Jones 1989b, p. 459), citing expense and uncertainty about their competitive necessity.

General Electric's factory automation division – which was earlier publicised as 'America's factory of the future supermarket' – has been quietly scaled down, after large losses. FMS and other computerised systems turned out to be difficult to sell as 'turnkey' products (Petre 1985). Even the FMS 'islands of automation' strategy has lost its earlier appeal (Baxter 1989). Market forecasts have been substantially scaled down (Boston Consulting Group 1985) and the gap between theoretical and achievable range flexibility has become received wisdom amongst both vendors and buyers of FMS (Zygmont 1986). However, the effect of this pessimism seems unlikely to stimulate attempts at more realistic policies to achieve flexibility of the 'economies of scope' kind. Indeed, FMS users such as Deere have gone back to Fordist essentials: cutting the time allowed to workers to reset machinery and redesigning components, so that automated lines have a more 'dedicated' character (Zygmont 1986, pp. 22, 24). For to gain a greater degree of qualitative or range flexibility requires social and political changes which are beyond the typical administrative and production paradigms of US managers.

The contrasts between three of the American FMS case studies underline this point.[6] At Locos management sought to win over the traditionally militant International Electrical Union local by paying higher wages in return for a replacement of specialised job classifications with flexible all-round roles. This arrangement was successful in getting the FMS though the early 'de-bugging' phase and in bringing down-time to levels that production managers could present as reasonable to their finance managers, and the outside world. However, management then rescinded the wage improvements on the grounds that these were special bonus payments for an exceptional period. The union, feeling that they had been duped so that the FMS could be made operational, then withdrew from the teamworking arrangements and crew members refused to do work that was outside their conventional job classifications. The union claimed, probably correctly, that the productivity of the FMS then declined.

This example suggests that it is management rather than unions who are incapable of the wholehearted commitment to the post-Taylorist flexibility in work roles that will complement the flexible scheduling potential of the FMSs. This is a conclusion which is confirmed by the contrast between two non-union FMS plants in the same southern state. At Bonsai the Japanese machine-tool manufacturer trained up local labour to generic FMS work roles. These included responsibility for remedial programming of the scheduling and part-program software. Management had an explicit policy of developing these skills in all FMS personnel. One hundred miles away at the US-owned Machinery plant, without any union to maintain conventional job classifications, management still retained specialised job grades in the FMS and refused to consider operator involvement in programming, saying, in classic Taylorist terminology, that 'it would not be functional'.

The inflexibility and dubious investment value of FMS in America seem, therefore, to be due to concepts of production efficiency and labour control in the managerial paradigms. Reprogramming to achieve more variety in systems made on the FMS, and responsiveness of the systems to new designs are rarely attempted because they mean more machine down-time and higher overhead costs while software and hardware are adapted; and normally the FMS was purchased not for these purposes but to reduce labour and inventory costs. But these priorities are not, as Jaikumar suggests, directly caused by a Taylorist mentality; in the sense in which that term has been defined here. The root cause is the production paradigm in which the new flexible manufacturing technology is seen as a means of achieving the efficiencies of Fordism, in the hitherto antediluvian sector of small-batch metalworking.

Taylorism *is* relevant when considered as a paradigm for the administration and control of labour. This certainly reinforces the Fordist production paradigm by maintaining occupational specialisation and low trust in production workers' decision-making powers: qualities manifested principally in the separation between 'managerial' programming responsibilities, and the mechanical-adjustment tasks of the work crews. Insofar as this is institutionalised in the contractual individualism of unionised plants, Kelley's (1989, p. 246) arguments against Marxist-inspired theories such as Braverman and Nobel are correct. *Pace* Kelley, workers' rights under the US collective bargaining system also contribute to narrowed work roles. Unilateral managerial strategies to deskill and subordinate are an understandable inference from US experiences. However, hierarchical job specialisation is not reducible to such an all-encompassing logic. As the previous chapters, and the case of the non-unionised Machinery plant show, a more fundamental underlying cause is the control and organisation of Taylorism as an administrative paradigm. This perspective is further reinforced by the defensive character of collectively bargained job classification systems, and the pursuit of Fordist efficiencies.

By and large this combination of Fordist computer-controlled technology – for lower cost production of small batches – with Taylorism is not interested in economies of scope nor in the benefits of flexible specialisation. So are these last two perspectives simply academic formulae without practical manufacturing feasibility? Or, alternatively, might such CIM technology as FMS be combined with one of these more flexible paradigms, where specific changes in the general conditions of Fordism and Taylorism allow? Evidence to answer these questions will accumulate in the succeeding chapters on contemporary metalworking automation in Japan and Europe. However, almost by chance, the American fieldwork provided a working example of the use of a FMS for something very close to flexible specialisation. Because it is the exception that explains how the North American rule might break down, the whole of the next chapter is devoted to an account of how, and why, the FMS at Alpha involved a break with Taylorist and Fordist practices and the establishment of a form of flexible specialisation.

7 An American deviant: FMS at Alpha

In industrial case study research it pays to keep an open mind. Many different aspects of a plant – its physical appearance, the interactions amongst employees, the signs and symbols that communicate information – can all convey clues about the character of its social organisation. Intent on getting accurate answers to highly specific questions even the best researchers can tend to disregard these various sources of indirect data. However, when notice is taken of these circumstantial signs and symbols – and if the questions asked don't force the interviewees into a limited choice of alternative answers – other social realities can be reconstructed: situations that might not have been envisaged, or are actually concealed by exclusive adherence to more formal methods.

So it was when I went to the Alpha plant. This is an old FMS; the second or third to be built in North America. Set up in 1975, Alpha was already seemingly well documented in business and research reports. Having read these reports and talked to the plant's operations manager I was mentally geared up for another bureaucratically organised FMS, like those emphasised in the preceding chapter. I was expecting it to be run to Fordist efficiency criteria, and eulogised by a junior ranking executive in business school and 'hi-tec' jargon. But as soon as the visit got under way my sub-conscious was bombarded with impressions that contradicted these kinds of expectation. In the language of social psychology I was experiencing 'cognitive dissonance'. It was only when, later, I realised that this was not the typical FMS of quasi-Fordist batch production, but a near-genuine case of flexible specialisation, that I resolved the dissonance between expectations and reality. Alpha's was a worker-controlled FMS run on flexible specialisation lines.

Because of the atypicality of Alpha, and because of the academic controversy about the practical realisability of flexible specialisation (see Pollert 1991; Wood 1989), the rest of this chapter is devoted to two stories. One is a brief account of my experiences at Alpha. The other gives the history and culture of Alpha's FMS, as told to me by men who have worked on it. On the basis of these accounts the reader can decide for herself, or himself, how far my verdict of a spontaneous flexible specialisation is justified. What should be especially borne in mind is that previous social science analyses of this FMS came to radically different conclusions about the nature of work and the status of the workgroup. After Harley Shaiken visited Alpha in 1982 he emphasised that: 'The human cost of freezing the worker out of the system is that most jobs on the FMS are robbed of autonomy and satisfaction' (Shaiken 1984, p. 150).

Shaiken bases this judgement on his interview with one of the FMS tool setters and upon an unpublished survey of the Alpha FMS workers' attitudes conducted by Blumberg and Gerwin. The interviewee refers to the combination of boredom and stress in the operators' jobs and a paradoxical conjunction of apathy with vigilance and responsiveness to deal with non-routine and critical problems. Shaiken highlights Blumberg and Gerwin's findings that: all eighteen FMS jobs were rated at below-average scores for autonomy, higher than average for stress, and all workers except two operators felt that they had skills that they were not able to use (Shaiken 1984, p. 151).

Yet while emphasising these degraded features of the FMS jobs at Alpha, Shaiken simultaneously claims that the Alpha FMS, like others, is far from meeting productivity expectations and depends upon the skilled interventions of the FMS workers. How can these two circumstances – deskilled lack of autonomy and dependence on skilled intervention – coexist? Why did the work crew emphasise lack of autonomy and lack of opportunity to exercise their skills? Was it because the work was intrinsically unskilled or some other reason; perhaps because the workers had been recruited with higher than average skills so that the FMS jobs seemed to them to have restricted skill requirements? The following account tries to answer these paradoxes and inconsistencies by presenting additional evidence from a research approach different from Blumberg and Gerwin's structured questionnaire method and Shaiken's descriptive job analysis perspective. It may well be that Alpha represents an alternative future for factory automation.

A citadel retaken

Alpha is in an industrial suburb of a large Mid-western manufacturing city. There is nothing about its grey concrete exterior to suggest that it is part of any computerised transformation of American industry. As I was led to the FMS area by the operations manager, the plant's exterior seemed old and worn. Inside, the buildings were grimy and dim. Expecting an antiseptic high-tech environment, and a meeting with some senior manager, I had dressed in a cream-coloured suit with matching shoes. As we tiptoed across a floor awash with industrial oil, I began to realise my mistake. That impression was confirmed when I was introduced to Morris Stone,[1] the FMS operations planner. Although his job was officially a 'managerial' position, Morris wore black jeans, a check lumberjack shirt, and heavy industrial boots caked with the oil through which I had just navigated.

In the FMS control room Morris lounged on the edge of a grimy desk while I perched, uncomfortably, on the edge of a chair of similar condition, struggling mentally to remind myself that less than ten years ago this installation had been a shiny, new, technological innovation. As we talked the impressions mounted of a high-technology island that had been gradually swamped by years of traditional industrial custom and practice. The general grime, casual arrangement of chairs and desks, greasy files and papers could have been located at any fifty-year-old steel mill, iron foundry, or metalbashing factory in the world. It was only towards the end of our interview that I realised that the Coca Cola dispenser was actually a mock fascia concealing the central computer. At one point, three grease-stained men in denims and overalls – the FMS crew – walked, unannounced, into the control room and proceeded to pull out coffee flasks and sandwiches for their lunch. Whatever managerial preserve had originally been intended for this control room every aspect of it now seemed to have been effectively colonised by the nitty-gritty of shopfloor life. Reminiscent of the Morlocks in H. G. Wells's *The Time Machine*, it was as though the underground toilers had re-emerged to reclaim their birthright.

It was only after several interviews, both inside and outside the factory – the latter because neither managers nor FMS workers wanted interruptions

– that the causes and implications of this surreal inversion of technological control, and its contrast with the deskilled subjection in Shaiken's account, could be pieced together. I cross-checked the lengthy interview accounts of Morris the FMS planner, Gene Clark an FMS operator, and Leslie Fraser the UAW shop steward and former FMS loader, with managers' statements and published sources. These accounts confirmed that an administrative and technical semi-autonomy from the rest of the plant had evolved. This development was bound up with an operational philosophy much closer to flexible specialisation than the neo-Fordism of most other US FMSs. Autonomy had been achieved without anyone, management or labour representatives, consciously planning or agitating for it. This is the story of how that autonomy – often advocated by FMS analysts – emerged; not from a strategic plan or technological design, but as a serendipitous outcome of continuous contests that make up the 'politics of the plant'.

Managerial perspectives and 'strategy'

The Alpha FMS was initially a joint venture between a machine-tool manufacturer, seeking more experience in developing computerised machining systems, and Alpha – which sought equipment to machine castings into housings for a new design of tractor. Operating flexibility was a major criterion in the investment choice because there was uncertainty about the market reception for the new model. If a dedicated transfer line had been purchased and sales had been low then the line would have been under-utilised. If it had turned out that the model had only a short market life then a transfer line would not have recouped its investment costs before being scrapped or refitted. On the other hand, 'stand-alone' machine-tools would not achieve such a high speed of throughput. The machines in the FMS could, if necessary, be operated on a stand-alone basis. Moreover, unlike a transfer line, the FMS could be reprogrammed and retooled to process different components in the event of fluctuations in demand for the new tractor, or its phasing out by the company.

In order to staff the FMS it was regarded as essential to circumvent some of the normal union controls on the allocation of workers. There was particular concern over the practice of 'bumping', whereby more senior workers – with the longest employment in the plant – could claim the new FMS jobs. The fear was that more senior workers might not make suitable workers for a job in the FMS. Management's view was that few of the existing machinists were the right age or temperament to help refine the largely untested technology for Alpha's new production needs. Thus a two-fold strategy was developed to isolate the organisation and staffing of the FMS from the rest of the plant. Firstly, managers presented the FMS to the union as a new project to be staffed by a new grade of worker, 'FMS qualifiers'. This ploy had the effect of removing the FMS jobs, at least temporarily, from the normal processes of transfer and promotion amongst existing grades. Secondly, having won freedom of selection by the regrading manoeuvre, most new workers for the FMS were recruited from the external labour market. The first operators came from a small local 'jobbing shop' – a firm that specialises in small sub-contract orders from larger clients. Over the next few years Alpha found that this was a useful source of skilled labour as the staff complement of the FMS was built up. When I visited, the tool setter and two of the operators, as well as Morris Stone, remained from the original jobbing shop recruits.

Other FMS-using firms have had the strategy of isolating the FMS from normal plant personnel policies and recruiting skilled workers from outside. At Caterpillar, for example, departmental managers have battled with both the union and their own personnel and job evaluation staff to create classifications separate from the existing transfer and promotion ladders (Stanovsky 1981).[2] However, at Alpha the strategy had unintended effects which eventually loosened managerial control over the FMS. The FMS-operator job classification was at the top of the plant's machinist pay scale. This fostered feelings of higher status amongst the FMS crew and a feeling that they had special responsibilities and skills that deserved acknowledgement by management. Furthermore the job shop environment, from which most of the crew came, emphasised individual ingenuity and close co-operation on specific orders. As a result, a group identity was imported into the FMS. This had practical advantages in coping with unexpected or difficult problems, but it also furthered the social cohesion of the FMS crew.

The success of the FMS varied according to the company's product strategy. Alpha is at a disadvantage compared with its larger competitors in the agricultural equipment industry. It lacks a large dealer network and the overall size to wage aggressive pricing and marketing campaigns. The new Alpha tractor of the late 1970s unfortunately coincided with a general downturn in farming revenues. This confirmed the wisdom of the FMS purchase. For it proved possible to fill the spare capacity caused by the low orders for the new tractor with the machining of components for other products; a mix that would not have been possible on a dedicated transfer line. On a later occasion, Alpha was launching another new model with a design change involving repositioned engine valves. The product was due to be launched at an agricultural exhibition but construction – originally assigned to the transfer line – was held up. To bring forward the machining of the demonstration models would have meant refitting, and thus closing down, the transfer line for other models. The FMS, and its staff, came to the rescue by remachining some of the castings for the new model, so that they could still be worked on by the existing set-up of the transfer line.

Managers felt that the FMS had contributed substantially to keeping Alpha as a successful niche producer in a market dominated by larger firms. However, there were also reports that the firm's financial department continued to be dissatisfied with the performance of the FMS. Until the 'rescue' of the exhibition model in 1982 this pressure had been pronounced. As Jaikumar (1984) has explained, cost controls on plant investments in the USA, measure successful performance in terms of reduced 'man hours' (*sic*) per operation, and the spreading of other overheads over long production runs – because these reduce resetting and hence down-time. Morris Stone, now officially part of management but spiritually and socially still part of the FMS staff, was scathing about the inappropriateness of these financial controls and the accountants' ignorance of the practical advantages of not trying to run the FMS in the manner of a transfer line to reduce labour and overhead costs.

The work-crew perspective

Throughout the accounts of Morris Stone, Gene Clark and Leslie Fraser there was an emphasis on the independence of the FMS operation. The first two also implied a measure of prestige and personal authority in their experiences. Yet the management faculty of the local university had done a study a few years before the interview which had emphasised the low morale and dissatisfactions of working on the FMS. The interpretation of Morris and Gene was that the academics had taken insufficient account of the fact that a new round of pay bargaining was about to begin. For the FMS crew – knowing the report would be read by management – did not want to convey the impression of contentment, and so undermine their case for pay increases. Work on the FMS, it was acknowledged, was demanding and intensive, but it was also stimulating; while, for the main grades of worker, it also gave considerable control and decision-making powers. The occupational classification is outlined in Figure 7.1, in descending order of status and reward; with numbers on each of the three shifts in parenthesis.

The system supervisor, or 'foreman' as the crew called him, was responsible for overall managerial authority in the FMS. In one early

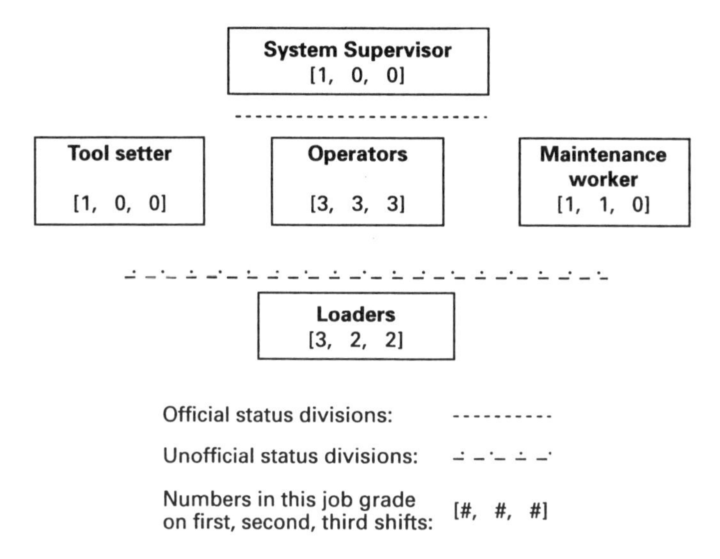

Figure 7.1 Occupational divisions on the Alpha FMS

report this post was even referred to as that of 'system manager'. However, this post was only filled properly on one shift. Because the FMS was operated on three shifts supervisory responsibility could sometimes be held by a foreman from one of the other machining areas. Because these non-specialist supervisors knew little of FMS methods and problems, such transfers were seen as often unsatisfactory by the operators, who resented being given orders by unknowledgeable superiors. The work crew had sufficient collective experience and *de facto* mutual transferability of tasks to be able to perform duties, such as work scheduling, that other firms would have jealously reserved for the system controller. Not only could they stand in for the system supervisor in his absence, but they could, and did, perform these tasks when he was present.

External supervisors often met with significant resistance from the FMS crew if they tried to exercise the kind of arbitrary 'right to manage' authority that prevailed in other areas of the plant. In the very early days 'Joe', a black worker also recruited from the jobbing shop, had gained considerable personal authority, by standing up to attempts by the rest of the shop management to subordinate the FMS operation to their procedures. He became a father figure, and role model, to the newer and younger recruits. Morris and Gene explained how, later, they would break any attempts by individual supervisors to reimpose conventional authority by non-cooperation and veiled threats to quit. The longer time passed by, the greater the force of these kinds of resistance; because the managers and engineers who had introduced the FMS moved out of the plant, or lost familiarity with its finer details. As the machines and equipment began to wear, and develop individual idiosyncrasies, so the knowledge necessary to correct and maintain the system's operational efficiency became particularistic and tacit. The Coca Cola mock-up on the central computer was highly symbolic. Dissatisfied with the delays and cost of awaiting service engineers for the computer, Morris and the others had sometimes improvised their own repairs for it.

Amongst the core group of workers – the tool setter, operators, and 'maintenance man' – there was considerable overlap of tasks and responsibilities. The crew's brusque take-over of the computer room for lunch, on my first day at the plant, was a normal event. Operators, and the tool setter, would customarily use the central computer to modify the schedules which controlled the routeing of the parts between the different work stations. Morris was not keen on operators changing aspects of part-programs such as the tool offsets, but admitted that *in extremis* setters and operators might have to do this. There was, however, a real barrier between the tasks of the loaders and the rest of the crew. The latter group were pariahs in terms of task sharing. The rest of the work crew thought that the monotonous and simple task of clamping the workpieces to the pallets was beneath their skills.

The loaders tended to lack previous experience and qualifications and their normal, subsequent career route was a machining or assembling job elsewhere in the plant rather than in the FMS. However, although the loading job involved a very narrow task and skill range their proper execution was vital to the functioning of the FMS. If the workpieces were not correctly clamped to the pallet, time-consuming readjustment, or even inaccurate machining – probably compounded as erroneously dimensioned workpieces continued on, unattended, through the FMS – could subsequently occur. The loaders, according to Leslie Fraser, had to be vigilant in looking out for mis-shaped castings and in avoiding over- or under-tightening of the clamps.

It was difficult to determine the exact events that had led to the autonomy and polyvalence. This was mainly because Morris Stone was now such a dominant figure, the FMS group's independence was bound up with his know-how and forceful personality. During one of my interviews with him, we were interrupted by a request for some urgent reprogramming for some other machines. Morris turned down the request curtly, saying that he was too busy discussing FMS issues with me. A college drop-out, Morris's employment in the jobbing shop, had given him basic machining skills. After a couple of years as an FMS operator, he had begun to take an interest in the programming aspects of the system. He took the software manuals home and studied them in his free time; helping out where possible, with programming tasks. Eventually the effectiveness of the system depended so much on his programming expertise that he was given an office and the job title of 'process planner'. However, informally, the FMS was seen as his main responsibility and, because of his experience, expertise and personality, he had most control. Yet, he regarded the FMS crew as self-contained. He also saw himself as still central to its operations, which were critical for Alpha's manufacturing processes. So Morris, in his jeans and industrial boots, acted as the FMS crew's champion within management.

Gene Clark, the tool setter, was prepared to defer to Morris's broader understanding of the systems and his greater software expertise. But this relative inferiority by no means implied that his own role was negligible. He regarded his earnings as substantial but fully justified. The FMS job meant not just greater stress – as Shaiken, and Blumberg and Gerwin claimed – but much higher degrees of analytical reasoning combined with conventional machining know-how. As 'operators' in some of the other FMSs studied also reported, error detection and even minor mechanical adjustments often presupposed a mental reconstruction of the processes in all three of the operating systems: the machining process, the part transfer processes and the system scheduling software. Because some adjustments and corrections could delay the entire system's operations, interventions had to be speedy and accurate. There was little room for errors and experimentation.

The operators, tool setter, maintenance worker and software schedulers tended to work together co-operatively. They swapped tasks, within the limits of experience and learning on-the-job, because the interdependence of the machines and the different systems created composite tasks. These were not easily sub-divided according to conventional occupational specialisation, without delay and complications. These operational exigencies were a necessary but not a sufficient condition for polyvalent work roles. In other FMSs similar pressures existed, but only at Bonsai, and for a brief period at Locos, had the operational requirements been accommodated by some task sharing. Moreover in both these cases it was the personnel policies and labour relations which created the space for whatever relaxation of job classifications had occurred. At Alpha the determining factors were a mixture of the market environment and informal social dynamics. An early need to change the product mix in the FMS combined with the accompanying sense of special identity and solidarity that developed amongst the FMS crew. The space for these developments was created by an element of tacit autonomy within the industrial relations procedures of the plant.

The industrial relations space

Leslie Fraser had been recruited into the loading job on the Alpha FMS from outside the firm, and had no previous machine shop experience. Later he became a shop steward for the UAW which bargained for all of the production employees. He no longer worked in the FMS but represented the workers in that area. In his account a combination of chance and pragmatic union tactics had led to the FMS crew being tacitly accepted as a semi-autonomous bargaining unit within the plant. At first the union had been concerned that management had effectively taken the FMS jobs out of the seniority system for bumping and promotion by the new classification of 'FMS qualifiers'. However, when it became clear that the unit would be an exception to the general industrial relations rules of the plant, union opposition slackened.

However, formal industrial relations peace did not guarantee real harmony. The solidarity and organisational integrity of the FMS crew began leading them into recurrent disputes with line managers. Because the crew were often successful in these disputes, the UAW stewards began allowing them to represent themselves in some cases that would normally have had union involvement. As Leslie Fraser explained, this required that the union 'turn a blind eye' to the fact that many disputed issues were being resolved in ways that ignored the procedures for grievance resolution, laid down in the union–management contract. On the day that I had tip-toed over the oil in the FMS area the crew had their own meeting with management to demand that the oil problem be solved for reasons of safety. Leslie Fraser was happy with this situation. For the FMS workers were, of course, all union members and, because they usually fought from a position of strength, they were unlikely to need to embroil the union organisation in their disputes. In this way the stewards could concentrate their time and resources on other areas of the plant where management were in a stronger position.

Moreover, many of the issues with which the FMS crew conflicted with management were often not legally within the sphere of union bargaining. In conformity with American labour law, and other union–management contracts, operating methods, production schedules and task allocation to staff are all 'management rights'. As we saw in chapter 5, practical exigencies mean shopfloor groups and plant unions can develop considerable indirect influence over these areas. However, they are not normally issues on which a plant union would feel able to mount a continuous and comprehensive challenge. Yet this is, in effect, what the FMS crew were doing by organising their tasks and the operation of the FMS largely according to their own judgements. The union's point of view seemed to be 'good luck, but don't expect us to handle these problems for you'.

Management–labour relations in the Alpha FMS therefore departed from many of the industrial relations principles of the Alpha–UAW contract, and US labour relations in general. The work crew had become a semi-autonomous bargaining unit, within the overall jurisdiction of the union. They made decisions which were outside of the normal sphere of 'bargainable' issues, and which were formally 'management rights'. Something had happened at Alpha which not only contradicted the diagnoses of critics of US labour relations – that the work group and its tasks have become segregated from the processes of worker representation – but seems to have reversed this separation.[3] Moreover, this had not come about because of a sophisticated managerial strategy of 'human resource management' to regain worker commitment by deformalising collective bargaining, as some IR experts are now predicting (Kochan, Katz and McKersie 1986). This situation at Alpha was largely an unanticipated consequence of management policies.

Conclusion

In a limited, but significant way, the Alpha FMS was a successful balance of factory technology with workshop social organisation; consistent with the prescriptions of the 'socio-technical' concepts of the Tavistock school of human relations (Trist et al. 1963). Alpha was consistent with Piore and Sabel's model in that there was flexible specialisation in the operation of the FMS on the basis of polyvalent control by most of the work crew. Yet what happened there is not likely to happen on a large scale in comparable US plants. The causes were largely a fortuitous combination of business and market conditions, managerial labour policies, and union reactions.

The FMS was initially purchased to cover a number of manufacturing options, and it was certainly successful in meeting deadlines and substituting for the inflexible transfer lines. For these reasons factory managers had to accept that there was some justification in not applying normal operating controls to the FMS. The company's niche position in the market for agricultural equipment predisposed it to be flexible, rather than striving for mass production and competitive tactics.

In other firms the isolation of the FMS from the plant's prevailing labour relations processes, for recruitment and grading, can be expected to enhance managerial control at the expense of the workforce's influence. This was certainly the case in a similar British FMS which is described in chapter 8. Yet at Alpha the recruitment of skilled machinists from the jobbing shop created a core of FMS workers who expected autonomy, were accustomed to co-operative working, and had the skills and capacity to undertake complex adjustments to the machinery and controls. Finally, the pragmatic tactics of the plant union in giving the FMS crew *carte blanche* for self-representation on particular disputes between themselves and line managers increased the crew's collective identity and solidarity.

Alpha is, in some respects, the exception that proves the rule. It was only because of specific market and business circumstances that management tolerated deviance and semi-autonomy as the price for the necessary flexibility. It was only because both management and the union could accept the FMS as a neutral zone in the trench warfare of contract bargaining that its workers gained some autonomy in the acquisition and exercise of skills. In this respect Alpha confirms the otherwise overarching influence of the US system of industrial relations on work roles and technology utilisation. A force whose influence most US analysts of technological change, with a few exceptions (see Zuboff 1988, pp. 403–4), fail to recognise as a source of restricted work roles and low-trust management.

Experiments with comparable autonomy in the Locos FMS in chapter 6, and on CNC machines, foundered on managerial doubts about deviations from the contractual individualism of corporate industrial relations (Fadem 1976). On the other hand, it has to be acknowledged that modifications to conventional industrial relations rules are occurring. The job classifications which specify occupational tasks are being broadened and intra-plant seniority rights, which govern mobility, are being loosened (Fadem 1986). It is doubtful, however, whether such official schemes as the Caterpillar reforms reported by Miller and O'Leary (1994) will relax 'top-down' controls over programming as many FMS-type schemes are introduced precisely in order to enhance those controls. With so much at stake both unions and managements are likely to make only limited concessions on job controls. In these respects the Alpha case is likely to remain an isolated vanguard.

However, Alpha does confirm that flexible specialisation is a practical operations strategy for firms who combine advanced technology with an appropriate delegation of responsibility to production workers. Should comparable conditions be possible in other plants then it is likely that the classic model of flexible specialisation will be realised. Multi-skilled workers having a high degree of operational control over reprogrammable production systems will make best use of these capabilities to make small to medium-length runs of a range of distinct products. However, this is not a rose-coloured picture of the management and work roles in the Alpha FMS. The Shaiken and Blumberg and Gerwin interpretations of what happened there have a substantial germ of truth. The work was potentially stressful but that was because it was often demanding and challenging; the borderline between these two states is very fine.

The FMS crew was certainly not one big happy family. Because there was no official policy of redistributing skills and responsibilities the shifts towards polyvalence that did take place were restricted by personal friendships, aptitudes and previous levels of experience. My interviews confirmed that the loaders were regarded as outsiders by the core group, confirming Blumberg and Gerwin's finding that this group and the operators accounted for nearly all the dissatisfaction with the job. The general finding that many skills could not be exercised is simply a statement of fact. Workers used to setting and operating conventional machine-tools cannot normally do so on computer-controlled machines. Because several of the crew had been all-round machinists in a jobbing shop – reliant upon manually controlled conventional machines – it was inevitable that there would be a gulf between those skills and the tasks on the FMS.

As other commentators have argued (Adler 1990; Hirschhorn 1984; Zuboff 1986), there are also new skills arising from the 'informatisation' (Zuboff) logic of computer integration and control. In metalworking batch production these also involve the analysis and modification of programs and diagnosis of automatic control faults. But, unlike Zuboff's paper-processing example there is not a complete transformation from manual-experiential skills to 'intellective' interpretation of new types of information. More frequently there is an oscillation between computer-mediated information modes of work and traditional manual-experiential techniques. Sometimes the latter may still need to displace the former. The one mode interacts with the other through the 'envisaging' skills mentioned in chapter 4. However, in their enquiries at Alpha, Shaiken, and Blumberg and Gerwin, did not ask about Alpha workers' exercise of these new skills. It is also likely that the reason that some of the six operators in the survey were unable fully to exercise their skills was precisely because task sharing was informally determined within the workgroup. It would, therefore, be subject to the intrinsic centrifugal social dynamics of work groups recorded as far back as the Bank Wiring

Observation Room in the Hawthorne Studies (Roethlisberger and Dickson 1964).

Yet, all of these shortcomings and dissatisfactions are part of a larger picture in which the exercise of broader skills and degrees of decision-making, for at least some of the workers, was gained at the expense of managerial control. In the opinion of Morris Stone and Gene Clark the academic reports of dissatisfaction were exaggerated. Morris Stone thought this was mainly because pay negotiations were in the offing. From the workers' point of view – knowing that the Blumberg and Gerwin report would be read by management – an impression of satisfaction would not have helped their claim. In addition, the emphasis in these reports on the constraints on job autonomy and skill usage reflects a finding from my interviews. There remained a tension within the firm between the striving for work role autonomy and flexible use of the system amongst the crew and some of the managers on the one hand, and on the other, the Taylorist orientation of line managers and Fordist efficiency assumptions of the financial staff.

It must be stressed that the main difference between the above account and the existing reports on Alpha is not just a claim that a full-blown flexible specialisation has won the day there. The difference lies in recognising the range of conditions in which the spontaneous development of flexible specialisation practices have both fuelled expectations and dissatisfactions; and also have formed the basis for the work crew's contests with the unreformed Taylorism and Fordism that surrounds them. At Alpha the FMS crew make good use of the magic broom, but their capability is far from being fully admired by the sorcerer.

8 Easy-peasy Japanesy: flexible automation in Japan

At a Tokyo subway station an automatic machine failed to dispense a ticket in return for my coins. As often happens in Japan, a small group formed to try to solve the problem. One Japanese began, in cross-cultural fashion, banging the machine with his fist. Immediately a panel in the machine opened. An angry human face appeared within the machine and, via an interpreter in the crowd, instructed me to hand over coins. The ticket duly appeared. For me this incident was highly comic, though not to my Japanese helpers, but also a symbol of Japanese mechanisation. Though processes are apparently automated, or 'high-tech', human assistance is ready to hand, at just the moment when help is needed.

Has Japan by-passed Fordism? Can the future of the factory be found in the shadows of Mount Fuji? If so, is this because of higher commitment to advanced technology or superior social organisation of work? Achieving standardised quality and efficient control of varying production schedules is a persistent problem in small-batch manufacturing. As previous chapters show, pre-existing and still-evolving labour relations institutions blunted and complicated Taylorist administrative controls of labour in the West. Variability of products and components has meant grudging dependence on human skills and decisions – limiting use of the Fordist techniques which gave greater predictability and consistency to mass production. The Alpha case shows CNC and FMS could blend more easily with small-batch manufacturing methods for 'flexible' forms of efficiency. Yet this route is largely neglected. Instead, despite contradictory practical and technical conditions, the lure of the Fordist paradigm persists.

Computer Integrated Manufacturing (CIM), specifically FMS, was seen as offering the prospect of mass production flows of work for small batches of different products. However, it has fallen short of expectations. As the preceding chapter showed, US users over-relied on the centralising

The expression 'Easy-peasy Japanesy' is used as a playground chant by English children.

capabilities of the technology, misunderstanding its flexible potential. The articulation of Taylorist occupational specialisation with contractual individualism in industrial relations, and the cultural hegemony of Fordism amongst managers, constrained technological success. In Western perceptions, Japan's manufacturing achievements are linked to an assumed superiority in the use of computerised production technology. For managers at a US FMS studied by Thomas 'modern' was conterminous with 'Japanese' (Thomas 1994, p. 86). One particular focus has been on Japan's apparent reconciliation of antagonisms previously deemed irreconcilable in the West:

- between efficiency and quality objectives;
- between technical complexity and human abilities; and
- between the organised interests of management and labour.

Do these apparent successes mean that many Japanese firms have by-passed the Taylorist and Fordist paradigms which have so far dominated Western operations (cf. Piore and Sabel 1984, pp. 217–20)? This chapter tackles the questions of why Japanese firms seem to have overcome the obstacles to computerised batch production, particularly the CIM orientation of FMS, and whether, as managerial writers like Jaikumar and Bessant (1991, p. 113) claim, they have shifted to a new production paradigm? Prevailing accounts of Japanese firms' recent industrial development suggest two major types of explanation:

1 A shift in management paradigms has let Japanese managers adapt labour and technology to post-Fordist methods.
2 The pre-existing network of industrial relations and employment institutions was already congenial to the methodology of computerised small-batch production.

A third possible variation on the latter explanation is that Western observers exaggerate the distinctiveness of the Japanese approach. Is the latter simply more successful Fordist revisions to conventional batch production, arising from a contingent combination of social and political conditions? As we analyse these possibilities we need to recognise that the difference between the first two views hinges on a question of the relative importance of either the superiority of practical management techniques, or broader social factors. Jaikumar has forcefully stated the first explanation by claiming that greater perspicacity and vision allow managers in Japan to adopt more appropriate forms of work organisation and hence greater productivity and flexible use of FMS.

> Most important, managers see FMS technology for what it is – flexible . . . not bound by outdated mass production assumptions, they view the challenge of flexible manufacturing as automating a job shop . . . the vision that leads to it is in human scale. No magic here – just an intelligent process of thinking through what new technology means for how work should be organised. (Jaikumar 1986, p. 72)

Whittaker's comparative Anglo-Japanese study of mainly CNC installations has the contrary emphasis. Unlike Jaikumar, he stresses their tandem involvement in the broader employment relationship, specifically training and individual wage rewards. While accepting a strategic role for Japanese managers in shaping work organisation, Whittaker also plays down the strategic autonomy of Japanese management in such initiatives, arguing instead that much broader historical and social contexts shape the managerial role, in both work organisation and the employment relationship. Yet, though occupational identity, training channels and general education are identified as relevant factors, their general salience is not systematically related to job control (Whittaker 1990, pp. 161–3).

To pin down more specifically any distinctively Japanese management of computerised machining, and especially advanced forms such as FMS, does require an assessment of management practices as compared to the influence of broader institutional forces. In the second half of this chapter my first-hand observations of the operational methods of Japanese FMSs will be linked to these managerial and social variables. First of all, however, we need to assess the more general accounts of Japanese manufacturing proficiency which, like the contrasting Jaikumar and Whittaker explanations, emphasise either managerial technique or cultural determinacy. This involves, firstly, the development of relevant institutions, the social trajectory, and then secondly, the Japanese adoption of computerised machining, the technological trajectory.

The social trajectory

The cultural approach is relativistic. It ascribes success mainly to the subsumption of capitalist productive criteria beneath the Japanese social context: enduring cultural standards and social practices. Its underlying argument is that shared values, and a sense of collective identity amongst managers and employees, minimise friction and conflict of interest, as well as maximise co-operation. As regards the second general approach, what we may call 'organisational imperatives', the social-institutional framework is secondary to the greater enthusiasm and rigour with which Japanese managers have adopted and adapted sound management techniques compared to the West.

Technical or societal determinants?

The earlier post-war accounts of Japanese manufacturing methods, written before Japanese firms were surpassing Western counterparts, tended to view Japanese forms of organisation and styles of management as pre-modern survivals and even impediments to 'modernisation'. Abegglen's 1958 account of *The Japanese Factory* established a benchmark by emphasising the distinctiveness of the *nenko* 'life-time' employment practice; associating this with long-standing societal values such as paternalism. By the time (1973) of Dore's classic – and more systematic – study of the differences between the factories of one Japanese and one British electrical machinery firm, the main issue was convergence and mutual borrowing between Japan and the West. Dore disaggregated *nenko* into the following components of a distinctive 'Japanese employment system':

- a seniority-plus-merit wage system...
- an intra-enterprise career system...
- enterprise training...
- enterprise unions...
- a high-level of enterprise welfare...
- the careful nurturing of enterprise consciousness... (Dore 1973, p. 264)

The resulting system 'seems to be more productive and better capable of delivering to all in it a steadily rising and better level of income' (Dore 1973, pp. 264, 274). Dore, like Abbeglen and others, still emphasised that key elements of the employment system derived largely from broader Confucian and group-loyalty values (Dore 1973, pp. 401–2, 473). The cultural solidarity between workers and managers is frequently described in terms such as 'paternalism' and 'familism'. For critics it is just this emphasis which is misleading:

> In striving to report what is most distinctive or unique about the Japanese factory they fail to keep what is most important in perspective. (Marsh and Mannari 1988, p. 336)

These critics argue that company performance results from the firm-specific form of universal organisational factors: rank, pay, job classification, seniority, education and the 'informational level – i.e. aggregate education, skill and responsibility scores – of the section in which one works' (Marsh and Mannari 1988, pp. 45, 332–8). This interpretation clearly overlaps Jaikumar's emphasis on managerial vision and expertise for higher FMS performance. It is also shared by other managerial writers who cite successful 'transplants' of Japanese operations into the West, to claim that what is distinctive and successful are sets of universally applicable organisational techniques (cf. Schonberger 1982). It is not possible to assess the wider progress of Japanese industry here but we can note, in passing, that a few British or North American branch plants no more prove that Japanese manufacturing is a bundle of exportable techniques than did the efficiency of the Indian railways signify the reproduction of nineteenth-century British capitalism in the sub-continent.

Our question is whether key societal context factors, such as Dore's 'employment system', and the typical Japanese batch production operation, combine to represent a break from Western batch production practice and Taylorist and Fordist attempts to rationalise this? To clarify the issue social institutions must also be related to Japanese manufacturing's general structure. For its 'dual system' of production and employment includes not only firms with branch plants in the West, but also various tiers of secondary employment and networks of smaller – usually sub-contracting – firms.

Work organisation and the broader employment system

The impact of Taylorism, as a hierarchical structure of work control and specialised work roles, was as limited in Japan as in Western Europe during the inter-war period. If they addressed this specific issue it would be difficult for advocates of trans-cultural managerial technique to explain why a methodology practised so widely, and to some effect in the USA, was marginal to most Japanese industry. Littler's synthesis of specialist sources, from the opposite perspective of cultural contextualism, argues that a rival indigenous approach based on fluid job boundaries, work groups and the continuing planning role of the *oyakata*, prevented adoption of key Taylorist features in this period (Littler 1983, pp. 149–50, 153–4, 157). Following Levine's interpretation (Levine 1965; Levine and Kawada 1980), Littler pinpoints this latter category, roughly comparable to the master craftsman of pre-industrial Britain, as the crucial figure in the changeover from pre industrial to industrial factory forms of work organisation. In this account enterprise managers manipulated workers' deference and dependence in the *oyakata–kokata* (master–client worker) relationship to maintain its traditions of fluid autonomous working, within a new framework of normative commitment to the enterprise hierarchy.

While logically plausible this incorporation thesis still raises more basic questions. Firstly, it does not explain why it was considered sufficient to leave the actual organisation of tasks – so fundamental to Taylorism – to the *oyakata* in their new guise as company supervisors. Secondly, there is the chronological problem that from the early 1930s to 1945 the politics and economics of totalitarian militarism clouded the direction of Japanese manufacturing. The government suppressed trade unions and worker militancy. Manufacturers became geared to producing for a protected war economy, without direct competition from the operational efficiency of Western firms. The officially backed Scientific Management movement of the 1920s regressed in the 1930s into a chauvinistic propaganda apparatus for authoritarian management (Nakase 1979, pp. 188–9). Thus, like Western Europe, the critical period for applying Taylorist methods in Japan was not the inter-war period examined by Littler, but the immediate post-war years.[1]

After all in Western Europe, as we saw in chapter 4, Britain and Italy thought post-war economic development necessitated the adoption of American management techniques, albeit through Marshall Plan auspices. Would not the same perceptions have some force in the war-shattered firms of Japan? There is a third critical question left unanswered by the *oyakata*-incorporation thesis: just what was so special about the work roles under this system that they have been so felicitously suitable for production in workshop, mass production, and now, seemingly, flexible manufacturing operations?

To answer these different questions we need, again, to clarify the purposes and character of Taylorist work reorganisation and specialisation in relation to Japanese preferences for shopfloor discretion. Remember Taylorism's aims: to specialise and control work tasks, normally by output-related pay, and to raise individual output, by-passing workers' cultural norms. For Taylor himself workers' collectively restricted output and workgroup customs' inefficiencies made this necessary. These Taylorist principles appeared specially suitable for improving the efficiency of the masses of inexperienced migrant workers thronging into US factories around the turn of the twentieth century. Italy shows the later appeal of Taylorism in Europe as a way of breaking the collective restrictions of shopfloor labour: and as a 'quick fix' for raising productivity amidst backward technology and workforce skills. Japan differs, I suggest, because social perceptions of identity and career made worker commitment less problematic. In metalworking, at least, a key factor was the status of the craftworker; as it evolved before the first Taylorisation attempts in the 1920s, and as diluted by industrial relations dynamics after 1945.

Whatever the character of workgroup activity in the inter-war years, the post-war period has witnessed a general increase in hierarchy rather than collectivist control of work: a development that complements the need to provide promotional movements for the enterprise-specific workers, but is distinct from Taylorist forms of hierarchy based on specialist occupational functions. Several of these developments only occurred in the late 1950s/early 1960s because the 1940s and 1950s were a period of instability in Japan's factories. Military defeat in 1945 and the Occupation forces' democratisation campaign facilitated a resurgence of trade unionism and worker militancy. These industrial relations developments influenced the framework for work organisation so their dynamics require separate comment.

Unions and the industrial relations trajectory

Japanese unions' main impact on the employment system came in the aftermath of the Second World War during the Allied – essentially US – military administration. After prohibition and suppression in the pre-war and Second World War periods, the Allied administration, constitutionally guaranteed and initially encouraged unionism until a surge of militancy swept Japan between 1945 and 1947. Politically discredited managers and industrialists initially lacked the confidence or resources to reorganise production on their own terms. Indeed, workers' take-over of the production process of factories for themselves was one popular union tactic (Ayusawa 1966). However, with the Cold War came American curbs on union powers – suspected sources of communist influence – and the rehabilitation of the managements of large firms. Managements gained the upper hand in a phase of bitter strikes until 1955. The resulting defeats for the unions led, similarly to Italy, to splits and schisms in their federal organisation. But unlike Italy this period of conflict and stabilisation institutionalised pay and employment policies which precluded direct Taylorist controls. The four key developments were:

- legitimation of industrial relations through enterprise-level unions;
- union aims refocused on to pay and employment security;
- surrender of the sphere of work organisation and job control on the shopfloor to management and supervisors;
- exclusion of Taylorist performance-pay control of work.

In the 1940s economic collapse, the break-up of firms, and the absolute poverty facing many ordinary workers, made employment security and an egalitarian pay system the main priorities of the newly emancipated and energised unions. Whatever the loyalty benefits of employment security to managers in recent years, during the conflictual 1940s and early 1950s these were concessions won by militant unions. From the 1950s wage inequalities were restored with the rightward shift in politics, and employers' recovery of confidence. But some employment security (Levine 1965, pp. 456–7) and seniority and needs-based, rather than performance-related pay (Gordon 1985) were also retained. Enterprise-based unionism as the dominant model reinforced priority for employment security – all non-managerial full-time employees of the firm were unionised (Nohara 1987, p. 17). The sources concur that unions' preoccupation with permanent employment – a tacit norm rather than a formal right – deflected concern for work organisation, which remains a managerial prerogative, albeit one exercised diffusely through supervisory hierarchies. Union priorities for pay linked to workers' age and cost of living, focused on seniority, but downgraded work organisation and methods. Unions' goals became the reward for work rather than its character (Kumazawa and Yamada 1989, pp. 122–3). However, their consolidation in enterprise bargaining precluded the Taylorist nexus of output-related pay based on task analysis.

Nor does enterprise unionism, and its neglect of work issues, signify uncontested managerial control. A paradox of Japanese enterprise unionism is that senior shopfloor workers, many of supervisory rank, are often union officers. Though this can increase managerial influence on union policies, it also directly exposes such representatives to day-to-day events at the point of production. Thus their occupational status increases their responsibility for responding to, and resolving, workers' associated grievances. Some reports show the supervisor as union officer playing a significant role in resolving, or representing workers' concerns to management (Inagami 1983). Having senior employees as representatives is also said to check the possible elitism and nepotism arising from having full-time stewards on special pay relationships.[2] Finally, Japanese workers' weak occupational identities are balanced by the wider solidarity of the unions' multi-occupational membership. In contrast to either de-unionised white-collar sections, or the dichotomous unionisation of white- and blue-collar workers in the USA and Britain, inclusion of both white- and blue-collar workers reduces potential competition and job boundaries between these groups (Koike 1987, p. 328), and hence management powers to divide and rule.

Craft, control and careers: the bureaucratic trajectory

In the West pre-industrial craft forms either atrophied through mechanisation and work rationalisation, or were consolidated into trade union identities. In Japan, the craftmaster–client worker relationship survived in a pre-industrial form until well into the Meiji period: i.e. the period of primary industrialisation. However, the strength of the traditional relationship varied considerably. In some instances it remained an effective means of passing on expertise and providing work: in the larger firms where the *oyabun*, or master, had sub-contractual responsibilities for work organisation and recruitment; or where the *oyabun* had his own workshop. But in other instances, and especially toward the end of this period, the traditional style of training by observation and job rotation degenerated into arbitrary restriction of the dependent workers to menial tasks.

However, the decisive survival from this arrangement was the expectation of upward career mobility from lengthy experience in specific jobs and tasks. Movement between jobs with different tasks became functionally equivalent to current Western forms of training. The way that the early schemes of formal instruction were used reinforced such practices. In the large firms they functioned to channel scarce recruits to higher technical and managerial posts, rather than to build up shopfloor skills to Western standards.[3] Between the wars career paths began to divide into today's dual structure (Dore 1973, pp. 398–9). Workers in large firms began to receive a mixture of formal instruction from enterprise trainers and on-the-job training through gradual movement between long spells in particular jobs (Levine and Kawada 1980, pp. 273–5, 290). After the Second World War, craft identity transmogrified into bureaucratic career in these firms, and was institutionalised in the form of the so-called 'Life Time Employment' or *nenko*. Yet in the small and medium firm, or secondary sector, traditions of mobility within and between firms persisted, initially under *oyabun* auspices. There the career goal became artisanal status, eventually as owner of one's own workshop.

Because jobs were already fragmenting within the *oyabun* phase Taylorism's job control and work specialisation aims became irrelevant. Also workers' commitments did not become craft or work group loyalties, because career success was seen as successful performance in specific jobs for movement and eventual promotion. With *nenko* in the inter war years, also came the main pay distinction: *joretsu* or 'experience' ratings. A series of status rankings synchronised pay-for-experience: 'promotion' came to derive from experience in a variety of tasks, based as much on length of service as actual performance. This system gave worker commitment and managerial power to allocate labour as required. Taylorist output-related pay 'would not have afforded management as great a measure of control over the work force and labour costs' (Levine 1965, p. 650). Moreover, like Europe, the most receptive time for transferring scientific management ideas into Japan was just after the Second World War, as part of what had become 'industrial engineering' in the USA. But industrial relations developments stifled this potential. Introduced within, and subordinate to, the *nenko* framework Taylorist techniques decomposed.

A related influence in larger firms was the expansion of supervisory roles. Japan has no strict equivalent to the Western occupational status of 'technician': an office-based, academically formed status mediating between higher engineers and production workers (Nohara 1987, p. 23). Instead, in the post-war period supervisors and senior production workers gained higher responsibilities for more advanced techniques; inspired by US industry's general foreman model, and promoted by legislative and training measures. In some sectors, a more technically responsible general supervisor grade replaced the *oyakata*-style of supervisory role. This *sagyocho* was ranked at, or about, the level of higher technicians or junior engineers, with some narrow supervision tasks assigned to new intermediate grades (Iwauchi 1969, p. 432).

In today's 'quality circles' workers may perform their own time-and-motion study (Kumazawa and Yamada 1990, pp. 115–17). However, concrete examples of the tasks actually organised around, and by, workgroups, as Littler claims, are hard to find in the literature. Task and decision sharing amongst discrete groups may have occurred in some sectors, firms or processes. But there is little evidence that collective job control is generally institutionalised. The oft-cited 'quality circles' do not direct the conduct of daily work tasks. Larger US and British plants organise work, or at least task allocation, mainly through occupational divisions. In Japan, however, the organisation is more likely to be based on individual workers' traits: 'age, seniority, level of education etc' (Nohara 1987, p. 16). Wood suggests that the higher ratio of supervisors in Japanese car plants over West German levels shows the supervisor's centrality in Japanese work organisation is incompatible with Western notions of autonomous and semi-autonomous workgroups (Wood 1989a, p. 452).

As in the case of the industrial machinery sector, below, the shopfloor workgroup's relative autonomy still stems largely from the expertise and authority of the supervisor and other senior workers. Unlike Western societies, work organisation in larger Japanese factories occurs within a wider and quite different structure of recruitment, transfer, promotion and reward: an alleged 'white collarisation' of most manual workers (Koike 1987). If a single characteristic defines workforce flexibility and commitment in such firms then surely this must be the institutionalised *individualism* of Japanese employment relations, rather than workgroup collectivism.

Economic and social dualism

The 'dualist' thesis, of an invidious divide between the employment arrangements in large firms and those in the small firm, has been applied to the Japanese context. It portrays small, labour-intensive firms as cushioning larger firms from the cost penalties they would otherwise incur from *nenko* and salary-type wage arrangements for their workers (Dore 1973, pp. 302–3, Twaalfhoven and Hattori 1982). 55.2 per cent of manufacturing enterprises have fewer than 100 employees – even though this figure excludes the size category of 1–3 employees – with roughly one-third of manufacturing small firms consisting of no more than four people (Patrick and Rohlen 1987). These figures suggest a robust and seemingly effective mode of employment.

Looked at from the point of view of workers' careers, however, it can be seen that the dual structure has come to serve a different function. Prior to mass higher education, an alternative form of industrial career to large firm *nenko* was a modified form of the *oyabun* career. Displaced from the larger firms these craft-style artisans, and their pre-industrial patronage of client and trainee workers, continued to dominate the small, workshop, sector. Client workers could often expect to set up their own workshop after an appropriate period in the employ of one or more *oyabun* artisans.

These firms' artisanal masters are not now, necessarily, the sole source of training for aspiring artisans. Yet their workshops have continued to act as a career route. Many of the frequent job changers in the small-firm sector are seeking the experience to start their own business eventually (Cole 1971, 1979; Patrick and Rohlen 1987). For metalworking employees in some districts these aspirations seem to have become more realistic over time. Employment growth amongst metal fabricating machine shops in the Tokyo area during 1968–78 was almost entirely due to 5,000 new 'small family enterprises' (Patrick and Rohlen 1987, p. 345). Some contemporary small firms may have unions and weaker forms of the larger firms' *nenko* system for their core workers (Koike 1987, pp. 103–4, 97–100). So small manufacturing firms' generally lower wages are not part of a complete dualism between high reward, job-secure primary employment and a secondary sphere of low-reward, unstable jobs. While such characteristics no doubt exist, in metalworking there is also a well-defined career route into independent artisans.

Large Japanese firms sub-contract relatively large proportions of their manufacturing to smaller sub-contractors; much more than by the large vertically integrated firms in a country like the UK. This is not just because of cheaper labour and other costs in a notional 'secondary' sector

(Thoburn and Takashima 1992, pp. 21–2, 130). At least in the secondary sector in metalworking, summarised in the discussion of technological change in the next section, there seem to be two main, and interrelated, reasons for small metalworking firms' vitality. Firstly, they provide the larger firms with reliable and quality sources for component parts and processes. Secondly, smaller firms' ensuing interest in skilled employees is bolstered by their workers' interest in skill acquisition, often as multi-task workers, because it facilitates an eventual career as an independent artisan. So while the Western form of craft identity never took root in industrial Japan, it remained important, as the informal basis of the artisanal career, in the small firm 'secondary' sector.

Summary

Insofar as labels are helpful then *meritocratic commitment* might be the most appropriate one for Japanese labour-management job control. Unlike the 'contractual individualism' in the USA, Japanese industrial relations processes do not extend into work organisation through the regulation of transfers and seniority. Japan is unlike Britain and Italy where the matching of task responsibilities and workers was regulated by 'craft job controls' or preoccupations with 'skill equalisation'. *Nenko* and the unions' focus on employment and reward have largely neutralised the relevance of tasks and skills as issues of collective grievance. In terms of the question posed at the start of this chapter, this system meant that occupational identities scarcely constrain management deployment of labour for operating efficiency. Skills are in short supply but tend to be acquired by individuals partly through actual or notional career paths.

Japanese manufacturing never developed organised craft work roles. The closest approximation was the gangs of client workers controlled by the master-artisan *oyabun* between the Meiji and First World War periods. But lack of systematic training and qualifications and the authority and monopoly of the *oyabun* precluded independent craftworkers in charge of their own methods. Prior to the spread of Taylorist principles and the consolidation of *nenko*, piecemeal work roles subject to rotation were most likely. It is a truism amongst Japan specialists, that skilled workers could not have pursued a craft identity and organisation until after the Second World War, as there were no effective national unions to affirm their status as ordinary employees. Yet after the struggles of the 1940s and 1950s the *de facto* compromises between labour and management created enterprise, not occupational, unions, and job security only for a category of permanent workers; and only on the basis of *nenko* career structures.

The relevance of Taylorism as either task specialisation or incentive pay systems was thus diminished. Yet no single feature of Japanese employment and management practices provided the complete alternative to Taylorism. The *oyabun* or *oyakata* arrangements of sectional or workgroup semi-autonomy could not alone have provided the supply of skills, nor the motivational commitment, to enhance productivity and handle new techniques. The refinement of *nenko* to include career paths and extra pay for new intermediate posts, and the expansion of the technical functions of supervisors into the *sagyocho* role, played a part. Union ascendancy over weak managements in the immediate post-Second World War period meant Taylorist methods were absorbed in the 1950s as discrete 'industrial engineering' techniques not as a complete system.

Neither rival explanation of cultural specificity or managerial excellence in organisational imperatives completely explains the importance of the deployment of workers and skills – and their combination with technology – for managing production operations. Managerial competence lacked the power to combine new manufacturing techniques and expertise without the depoliticisation of work processes resulting from meritocratic collectivism. Yet the latter was not a system that managers chose and constructed by themselves. On the other hand, the enduring legacy of Japanese cultural institutions cannot adequately explain the strengths of this arrangement. Lack of occupational status and organisation also make it difficult to accept the suggestion that Japanese workers' roles amount to craft practices (Piore and Sabel 1984, p. 220). The case studies of flexible production, addressed in the third section below, reveal whether the actual tasks of working with new technologies deserve a craft label.

The changes in the relevant institutions between the *oyakata* and contemporary periods have come through the economic pressure of skill shortages and the political pressures from labour, rather than through perspicacious managerial vision. Contrarily, as we shall see in the reports on Japanese FMSs, the institutions of the employment system also have flaws, which contribute to the distinctively Japanese use of this technology. Prior to this evidence, however, it is necessary to review the pattern of Fordist automation in Japan.

The technological trajectory

After the militarisation of manufacturing ended in 1945, and the social conflicts and managerial restructuring of the late 1940s, the technical reorganisation of Japan's dislocated industry began in the late 1950s. English-language sources suggest that Fordist-style industrial engineering was the main thrust of this initial technical modernisation. The spread of assembly-line methods, process controls, and standardisation of production schedules mirrored Marshall Plan aims in Western Europe. Similarly, the main recipient of these methods and machines was not small-batch metalworking industry. At least on the evidence of employee transfers and job redesigns, chemicals, shipbuilding, metal-forming, automobiles and electrical engineering were the main takers (Iwauchi 1969).

Industrial organisation and manufacturing strategy

As in Europe the spread of more automatic techniques and equipment in metalworking was limited by the typically small production runs for products. In Japan this factor, in key sub-sectors such as industrial machinery, was bound up with the small size of firms. While a dualist structure is of some relevance in these sectors its critical importance is not the sharpness of the divide between large, capital-rich primary firms and small, under-resourced secondary ones. The more important structural factor is the absence of a sharp division between large and small firms and the dynamic growth potential that some of the latter came to exhibit in more recent decades. A more detailed assessment of technological change in recent years is facilitated by focusing on the better researched and reported – in European languages – industrial machinery sub-sector. It also provides a backdrop to my first-hand experience of their FMSs described in the next section.

David Friedman has meticulously shown that the Japanese machine-tool industry grew despite resisting the politically inspired rationalisation of products and enterprise structure for Fordist production dimensions and technology. This contest goes back to the militarist–nationalist politics of the 1930s when Fascist-influenced technocrats and the military sought to build up Japan's indigenous machine-making capability through concentration of the enterprise structure. Both the giant *zaibatsu* monopoly firms and the small and medium-sized enterprises (SMEs) perceived these policies as contrary to their interests. Throughout the 1930s and 1940s the government took measures to get larger enterprise units and economies of scale – and hence Fordist production of standardised machines using interchangeable parts – in order to enhance the manufacture of military equipment. These all failed to overcome the resistance of the businesses within the sector which entailed both security from foreign competition and a stagnation of technical development (Friedman 1988, pp. 39–70).[4]

The post-war modernisation programmes resembled a replay of this pre-war rationalisation saga. The post-war champion of Fordism in the machine-making sectors was the Ministry for International Trade and Industry (MITI). From the 1950s onwards MITI was the architect and promoter of successive schemes to convert the heterogeneous product and enterprise mix of the Japanese machine-tool industry into fewer, larger firms, each making a narrower range of specialised machines. These schemes also failed because many producers preferred independence and informal relationships with sub-contracting firms to MITI's proposed cartels and mergers. Indeed, Friedman shows that in some ways the industry 'groups', set up by the firms to appease MITI demands for collaboration and specialisation, assisted smaller firms by transferring know-how and technology (Friedman 1988, p. 102). To date, the largest firms, but not necessarily operations, in this sector tend to be those with machine-making as one branch of multiple operations, while SMEs make only machine-tools (Nohara 1987).

Computerisation – diffusion of NC

This preponderance of small and medium-sized firms and diverse products limited the scope for dedicated automation in Japanese machine-making firms. Even those that produced in large batches tended to sub-contract to more labour-intensive 'secondary-sector' plants. When computerised automation – NC and robotics – was adopted from the 1970s onwards, its use was conditioned by this industrial structure; as well as by the work roles and employment relations described in the previous sections. Micro-electronic technology, particularly CNC, has bifurcated machine-producing firms into a paradoxical interrelationship. On the one hand, are a few large and medium-sized makers of more standardised machines and, on the other hand, a more heterogeneous mass of SMEs. These sub-contract to the larger firms using the flexibility of CNC to make small batches of products and components, or diverse orders.

Friedman sums up the relevant difference between Japan and the USA by the following formula. US machine-tool firms use batch techniques to make the special-purpose machines for mass production user firms. In Japan the machine-tools are made as standardised products for flexible general-purpose use by the more typical batch production firms. The larger, independent, specialist machine-tool firms have come to concentrate on large-scale production of standardised NC-type products. This strategy is assisted by extensive sub-contracting of detailed machining and sub-assembling (Nohara 1987). All of these developments must be seen in the context of an increasingly competitive domestic environment in the 1970s and early 1980s. Then non-specialist firms entered the market, attracted by the new sales opportunities arising from the simplification of the electronic control components. So competition on price, and reduction of production costs, has become crucial. Yet, customer firms still expect some variety to suit their specific needs (Nomura 1990). What possibilities for work and factory organisation followed from these newer technologies?

Studies of work with computerised machining

It is a highly conditional, quantitative proposition, but if other metalworking sectors have adopted similar product and production approaches, then the NC stage has meant greater in-house specialisation on high-value, precision machining. Also, most machining operations have been increasingly sub-contracted to smaller, often artisanal, firms; precisely because the latter's technical and productive capacity has been raised, at least partly, by adopting CNC. Japanese government figures show between 10 per cent and 29 per cent of small (less than 50 employees) with some 'electronic machines' (Nohara 1987, p. 14). Sub-contracting is also important because its artisanal base is at least as likely to retain, and reproduce, the all-round metalworking skills, which the larger firms' *nenko* meritocracy and bureaucratic work organisation tends to minimise.[5]

Computer numerical control

A detailed survey of work organisation of nine Japanese plants was undertaken by Whittaker, who compared these with British practices. As well as variations by size of plant, he found that more of the programming tasks were undertaken by the Japanese machine operators, especially in the smaller, artisanal, firms where 'programming was seen as a natural part of the operator's job'. Compared to British operations, the Japanese operators undertook a wider range of ancillary tasks, with responsibility for various machines, but had less training, and moved between functions more to suit perceived production priorities than career development (Whittaker 1990, pp. 154, 143, 114–37, 61–62). He sees enterprise policies as based on, if not induced by, socio-historical forces: the distinctive employment system, enterprise unionism, and the more potent cultural and normative resources which Japanese managements can invoke (Whittaker 1990, pp. 44–52).

On the whole Whittaker identifies a more 'technical' approach in his Japanese plants: with unattended machining, preference for computing skills rather than mechanical ones, and a general expectation that automation must eliminate costly labour inputs. However, each of these generalisations has to be tempered by key differences between artisanal, small and larger firms. In artisanal firms, for example, mechanically inexperienced operators worked CNC machines and all operators were expected to do some programming, unlike the tendency to prefer specialist programmers in the larger firms (Whittaker 1990, pp. 146, 157–8). If Japan's use of such advanced technology as CNC – some of these firms were also adding elements of FMS – does differ qualitatively from Western practice, then his account suggests that key elements in this success are: minimal formal training; wider task duties for operators; more managerial control of worker allocation; and labour displacement as a priority. However, despite the meticulous detail provided for chosen variables, Whittaker's study is less concerned with the overall organisational and product-related context of the Japanese plants.

Flexible manufacturing systems

While Whittaker sees Japanese factory automation distinguished from Western approaches by institutions and social practices, Jaikumar's FMS survey pinpoints managers' different national outlooks, perceptions and competences. He sees the management of skills as the specific cause of more varied and productive usage of these systems. Managers' perspectives are 'not bound by outdated mass-production assumptions' and allow:

- employment of more college-trained engineers;
- higher spread of CNC experience amongst the workforce;
- FMS project teams' involvement far into operating phases;
- writing and changing of schedules and part-programs by operators; who are
- 'highly skilled engineers with multi-functional responsibilities'; which
- 'work best in small teams'.

In broad outline this account usefully contrasts with American FMS approaches. Yet internal inconsistencies make it, at best, no more than a rough guide to key details of Japanese practices. FMS is specifically meant for batch producers, thus a strict 'mass production mentality' must be unlikely amongst such managers. More basically, the emphasis on manufacturing managers' strategic perspectives, as promoting and harnessing these factors, overstates managers' role. A more general caveat to the Jaikumar account is that Japanese managers can indeed deploy labour more effectively to exploit technology, but the depth of the labour skills is much less profound than he, and other commentators, perceive. Moreover, his account neglects the largely inherited nature of these conditions: have managers much choice to use labour in some other way?

For Jaikumar the innovative and responsive style of the Japanese FMSs represents automated 'job shops'. The problem with this analogy is that several machine-tool firms in his sample are recorded by other reports, as above, as seeking more standardised large-batch production. It also seems inconsistent to portray qualified engineers as a major source of FMS versatility, if shopfloor workers rewrite programs. Also, as is shown below, though Japanese firms recruit many high school and college graduates into the shopfloor, these need not be qualified engineers. Whittaker's finding of minimal engineering training of production workers may not be universal, but it shows that the logic of Japan's employment system contradicts systematic training for workers at operator level. More crucially, for skill deployment, inputs from high-skilled workers are kept to an economic minimum.

FMS case studies

Scope of the study

Proper clarification of the competing factors identified in the surveys of Whittaker and Jaikumar requires specifically designed observations and detailed case study interviews. My study lacked the resources for such detailed analysis. Nor will competence in the Japanese language give a Western investigator the key cultural prerequisites to understand fully the relevant social processes (Cole 1971, p. 51). My visits to Japanese plants, interviews with production managers and observation of FMS operations, amount only to abbreviated case studies. Four industrial machinery firms using FMS were chosen for their considerable FMS experience and because they operated up-to-date technology in different ways. As they use FMS in order to show off some of their products this choice of machinery firms is not necessarily typical of FMS users in general. But such firms often have more experience of cybernation. They also yield another kind of data: on the take-up and use of FMS by other types of user firms.[6]

My four FMS users were independent producers of industrial machinery. The two most automated firms, Sumo, and Yamanote, both employ 1,200 workers. Their FMSs were integrated with other computer functions and tended towards large volume production. The other two made medium-sized batches. The medium-volume producers, Bonsai and Karaoke, had 1,000 and 1,700 employees respectively.[7] Each firm had at least two plants; and except for Karaoke had the FMS at the newest site. Industrial relations seemed to have little influence on cybernation. Bonsai had no union. Only

Karaoke admitted formal consultation with unions and, despite a two-day strike over redundancies in 1976, consultation was limited to union representation on a works council. The union's only influence on the staffing of the FMS was handling subsequent grievances of individual workers. However, its latent power was shown after my visit. Union representatives helped to vote the firm's paternalistic founder out of the presidency.

Production policies

Unlike Jaikumar's post-Fordist descriptions, direct cost savings, not qualitative advantages of versatility, were the most often cited reasons for using FMS. At Sumo the three main aims were: maximum plant utilisation (24-hour days); savings on inventory and labour; and shortened lead times between designs/orders and final production. Their FMS was part of an overall strategy to create new investment funds from cost savings. Consistent, however, with Friedman's model of mass production for flexible usage, these aims differed from Sumo's clients' reasons for *buying* FMSs. These were to balance both cost saving advantages and gains in quality and flexibility.

Yamanote uses its FMS to make components for robots and CNC machines. It has an international reputation for its levels of automation – the 1980s' business press regularly cited its 'unmanned' night shift. Yamanote managers valued FMS for its contribution to 'total costs'; i.e. 24-hour operation, reduced labour costs and higher productivity. Like Sumo, cost savings were important at Yamanote for generating funds for further production plant investments. Bonsai similarly cited overall costs and higher machine productivity as advantages. However, for Karaoke the in-house aims of the FMS were difficult to separate from its demonstration functions for potential customers. The company had sold ten FMSs in Japan and one to a US firm. Sumo and Yamanote had the most focused FMS use, in terms of products made. Batches of up to 1,000 items passed through the Yamanote FMS. The Bonsai and Karaoke FMSs made a wider variety of components and machines: various parts for machining centres and CNC and conventional lathes at Bonsai, with 95 different components for various machine-tool products, in batches of between 10 and 20 on Karaoke's FMS. Variety is still the spice of life at Karaoke, which in its publicity emphasises the multiple combinations of machinery provided by its six different forms of machining systems.

These cases are only a small percentage of Japanese FMSs, but broader surveys in technical journals do show some division between more highly automated/larger batch users, and FMSs with less computer integration making more varied products (cf. *American Machinist*, February 1985, pp. 83–8; and Hartley 1984). So Jaikumar's picture of constantly innovative FMS users may be exaggerated – picking up only aggregate differences between ultra-conservative US users and Japanese counterparts. Some Japanese FMSs make a wider product variety, even though their level of computerised integration is lower, continuing previous policies of product variety and smaller batches. Japanese customer firms prefer machines tailored to their own specific requirements; some modify vendors' machine-tools themselves. This limits volume production opportunities (Marsh and Mannari 1988; Nomura 1990; Jones interviews 1985), but there is a modified Fordism: larger overall batches of standardised products involving less resetting, labour and machine down-time, according to Nohara. So many firms – like Nomura's machine-tool case – make standardised modules for the specific combinations customers demand to square the circle of large batch vs. product customisation (Nomura 1990, pp. 5–6).

Workers and work roles

Whittaker and Jaikumar differ on the causes of the Japanese practices, but both stress the importance of college-trained engineers for programming CNC and FMS (Whittaker 1990, p. 150; Jaikumar 1986, p. 71). If, furthermore, FMSs are introduced so as to reduce shopfloor operators' inputs, how can their skills still be so important? What exactly do FMS production workers do and who are they? Jaikumar calls them 'highly skilled engineers with multi-functional responsibilities'. Yet in Whittaker's sample of CNC firms – which included some proto-FMS plants – levels of training and experience were lower than in comparable British firms. The cautiousness of production managers, and the former predominance of craft training and responsibilities in Britain, partly explains this relative deficiency; but it hardly conforms to Jaikumar's 'highly skilled engineers'. Moreover, if operator responsibilities are important, then the crucial element is whether operators program. Jaikumar says they write and modify schedule and part-programs. Yet in Whittaker's CNC sample arrangements varied from firm to firm. Operators in the larger factories were more likely to edit and prove than do initial programming (Whittaker 1990, p. 146). Are FMS operators similarly restricted?

Significantly, in Nomura's FMS the engineers in the production engineering department did most programming; but some of the scheduling '. . . was transferred from Production Engineering . . .' to the '. . . group leader and some of the older workers' though 'Scheduling is complicated and only a few workers are able to do it' (Nomura 1990, p. 8). Co-operative working, rather than the level of engineering expertise, was also notable in my four cases. As in Whittaker's and Nomura's firms any individual operator would be expected to do mechanical adjustment and diagnosis tasks. In other words, 'horizontal' specialisation was rare. But the procedures for distributing skills generally limited operators' higher-level tasks, such as scheduling. Although not formally barred from taking on programming tasks as in the USA, these would not be expected of operators.

For example, at Bonsai it would normally be the FMS supervisor, or a programmer from the central computer department (five staff) who would change programs. Although, if these were unavailable, operators were able to take over. Operators were allocated to the FMS either because of their machining experience, or because there were more operators than work on their section. One of Bonsai's FMS operators we interviewed thought that the FMS tasks were not very different from his previous jobs during eight years at Bonsai. A sign at this plant exhorted workers to: 'Decrease human ability to increase productivity'. Even more restrictive roles applied at the quasi-mass production Yamanote. Schedules were also fixed and the foreman or production engineer, rather than the operators, made changes. The FMS here only ran ten main part-programs.

Yet at the other large-batch plant, Sumo, the range of thirty types of part-programs and many different workpieces probably influenced delegation of part-programming modifications to operators. At Sumo the shopfloor group also contributed to programming the schedules, and operators could modify them if the supervisor was absent. Karaoke was different again. Like British craft practices in some respects, there seemed less centralisation of decisions. The plant union was militant on some issues and a rare grade of skilled workers, 'cutting specialists', was set apart from assembler and operator grades. Only the highest rated operators and cutting specialists joined the FMS. They made some of the ninety-five part-programs, and helped make and modify the schedule programs as necessary. In all firms the only specific engineering training for production staff and supervisors was in-house company courses. For FMS tasks, most training was on-the-job.

This admittedly small sample is mainly relevant to the industrial machinery sectors, but it clarifies three aspects of computerised production work in Japan. Firstly, the key divisions between operating and programming tasks vary, partly with product ranges, and batch sizes, and partly with level and styles of managerial controls over labour. Secondly, the delegation of more responsible duties, within this spread of tasks, is informal and linked to operating exigencies. There is no general principle of operator involvement or team co-operation. FMS supervisors make most high-level shopfloor decisions – such as modifying the schedule. Thirdly, and unlike Jaikumar's account, workers' contribution to flexibility and productivity stems not from more detailed qualifications as 'highly skilled engineers', nor high numbers of college-trained engineers, nor collective decision-making, but from production teams' responsiveness and capabilities. These qualities are linked to managers' power to combine just enough skills, aptitudes and co-operation in workgroups – no more, no less.

Summary and context

So FMS use in Japan does not, in general, contradict Fordist production priorities. Compensating flexibility within these priorities arises from the distinctive features of the social organisation of the relevant human inputs – notably the absence of occupational identities in the larger firms and the weakness of Tayloristic controls. However, the basic sources of this organisation are neither a superior battery of management techniques, nor the expression of endemic cultural principles. More important is a socio-economic complex of institutions. Both firms' investment rationale, and the composition of the product, clearly signified that FMS use was not necessarily for varied types of components by innovative reprogramming, but to reduce labour and other costs, to increase overall productivity mainly for standard product lines.

This was not, of course, single-product mass manufacture like a conventional automobile plant; but neither was it Jaikumar's innovative, diverse product-range scenario, nor a case of flexible specialisation. If pigeon-holing is helpful, then most Japanese FMS users in the industrial machinery sectors are in a similar, but more efficient, 'neo-Fordist' category to their US counterparts. Significantly, for example, the small, non-FMS, maker of industrial presses planned to get an FMS to pursue what they themselves termed 'mass production' techniques. There are, no doubt, flexible small-batch innovative users of high technology; but these are unlikely to be FMS users, nor be large and medium-sized, primary sector, firms. Nomura, and Nohara (1987, pp. 9–14), suggest that larger firms' recourse to sub-contracting partly facilitates their use of FMS. Nomura's case study firm sub-contracted simple and/or labour-intensive work. Likewise, about 50 per cent of mechanical operations was sub-contracted in the firms I studied. The more aggregate data of Jaikumar may reflect conflicting and changing aims for Japanese FMSs.

The most diverse product range was on Karaoke's FMS, but the president wanted to use it to return previously sub-contracted work into the factory.

The two key points are, firstly, that Jaikumar's account of Japan's FMS, like other 'organisational imperatives' explanations of its manufacturing, over-estimates the importance of the plant managers' organisational choices in recruiting, training and distributing skills in relation to tasks. These are crucial to Japanese use of computerised production, but they stem largely from institutional arrangements – mainly non-Taylorist labour controls – that independently pre-exist particular management decisions. Managers at plants which vary from the micro-meritocracy model – as at Bonsai – must still adapt to those circumstances. The second main point is that Jaikumar's portrayal resembles some versions of the flexible specialisation thesis in exaggerating the compatibility of labour arrangements with flexible, post-Fordist production. Some Japanese batch producers use local versions of the employment system to optimise the efficiency of variable batch production. Yet significant other cases, even with more propitious versions, use labour and technological resources to pursue or adapt Fordist principles.

Conclusion

Unlike Britain and the USA, Japan's combination of industrial relations, the *nenko* system and dualism in jobs and industrial structure can supply skills to different production paradigms. Remember how Britain's hybrid pattern of union controls and Tayloristic task regulation helped maintain significant inputs from craft-status workers into the planning of machining and, later, CNC programming. By contrast, the craft element in US unionism was less significant in shaping work roles. US firms achieved more centralised, hierarchic controls over work planning, and subsequently CNC and FMS machining. The US collective bargaining system's contractual individualism in job classifications, controlled the supply of skills. In Japan, by contrast, craft-style identities and work roles were displaced into the artisanal sector but industrial relations institutions have not regulated the supply of labour skills.

In Japanese 'primary' sector firms workers' interests are not linked to occupational identities; which need not mean, *pace* Dore, workers necessarily see every management decision as in their interest. Pay is not directly linked to occupational grade and task responsibilities, but large proportions of the complex wage package come from proven performance in the exercise of skills and responsibilities (Whitehill 1991; Whittaker 1990, pp. 66–9). This is an incentive to work in jobs with more scope to display competence; though the individualisation of wages does not mean that downward transfer into a job with lower skill or responsibility results in wage reductions. The importance of promotion to the performance and ability component of the wage, and the detachment of grades from hierarchical authority, means a tendency to have several levels of shopfloor grade between the routine operator and the highest supervisor (Nomura 1990, pp. 13–15). Karaoke, for example, had three grades of supervisor between the workgroup (less than ten workers) and the section manager.

Pay and promotion provisions dovetail with the system of on-the-job training. The merit element in the wage may underpin the workers' co-operativeness – encouraging sharing of skills and working knowledge. The workers' chance to learn new task skills can assist pay and career progression, so the workers' aspirations will drive learning on-the-job; part of Koike's 'white-collarisation' of production workers. Normally, costly formal training programmes, or the external labour market are unnecessary. This system facilitates allocation of task skills to new developments such as FMS; with recent trends of full-time education extending to age eighteen reinforcing it.[8] The actual extent of transfers and mobility may be low, as Whittaker says, but the normative expectation is that any worker – within bounds of aptitude and competence – adopts new tasks. Sometimes the range may be 'horizontal' across mechanical adjustment and diagnosis, as at Yamanote; but it may be upwards into 'higher' programming skills, as at Karaoke. The deployability of labour skills constrains FMS operations in Western firms; but the Japanese model has limitations.

Precisely because the deployment and application of skills hinges so much on the *nenko* system, disruptions in the latter – such as a perceived decline in career promotions – could change workers' support for task co-operation and job mobility. Assumptions of growth and of the greater adaptability of young workers have led Japanese industrial machinery firms to place high school, even university, graduates, into shopfloor positions. At Yamanote and Bonsai, for example, the average age in the factory was 28 and 29. Changes such as intensified automation or contractions in product demand may sour flexible and co-operative orientations, as a large 'bulge' of similarly qualified and aged employees face reduced career prospects (Whittaker 1990, p. 49). Already Karaoke's management were anxious that younger workers desired 'too much democracies'.

Managers might avoid such dangers by less reliance on worker commitment through more complex automation. But such a switch to the American way may be even less useful than in the USA. The legacy of the current work practices could plague expanded CIM. Karaoke management complained that the informal work roles which keep procedures in the heads of the workers, rather than on paper records, hamper formalising procedures for computerisation (cf. also Haruo 1992, p. 20). If firms sought flexible specialisation production then supervisors' skills and co-operation amongst unequally qualified workers may not be enough. Flexible specialisation needs knowledge in depth in each individual worker to solve problems promptly. Japanese firms seem more likely to continue providing skills to computerised production in a human parallel to the famous Toyota 'just-in-time' parts supply system: discretionary rotation and on-the-job training gives workers just enough skill for an adequate production team: human *kan-ban.*

Computerised automation succeeds in Japan because the institutions of the employment system allow deployment of sufficient human expertise to optimise technologies such as FMS. Avoidance of Taylorist work organisation helps this capacity; but it does not mean a rejection of Fordism. Japanese FMSs apply Fordism more successfully to small-batch production than Western counterparts because Taylorism is limited. Yet Japanese neo-Fordism may be nearing its limits. More standardised, pre-planned production through greater cybernation would probably weaken the organisational and social conditions on which its present success is based.

9 Revolution from above: FMS in Britain

The 1980s were the heyday of Thatcherite 'market forces' policies for British industry. Yet, for much of the period, the Conservative government's Department of Industry orchestrated FMS and similar advanced computerised technology investments in at least thirty-nine large and medium-sized firms. Even the Iron Lady's free-market armour-plating was seemingly vulnerable to the lure of flexible technology. This state-led cybernation of factories developed from the avowedly interventionist industrial policies of the preceding Labour Party government. The common theme was British firms' weakness in international manufacturing competition. Strong and rapid technological innovations could reduce the lead of other countries in conventional and new technological capabilities.

Compared to other countries surveyed in this book it is ironic that it is in free-market Britain that central state direction has been most responsible for introduction of FMS and related technologies. As Friedman shows, and my own interviews confirmed, the most that MITI did was to facilitate adoption of computerised production amongst already innovative Japanese firms. In the USA the Department of Defense and the US Air Force were prodigious influences on advanced schemes for computer integrated systems, but FMS adoption was largely the result of the suppliers' exploitation of the Fordist, 'workerless factory', prejudices of manufacturing managers. In Italy, as well, the state has played a minor role – limited mainly to tax incentives – in the modest amount of FMS adoption that has occurred (Baglioni 1986).

The Thatcherite project for economic revival and social reconstruction has often been dubbed a 'revolution from above'. It sought a transfusion of market forces into the British economy which the government would enforce if businesses could not themselves accept this treatment. For most of the 1980s top managers reproduced this principle. Injections of market forces inside firms – redundancies, factory closures, new staffing levels – were imposed from the top; despite – though by no means always (see Rose and Jones 1985; Jones and Rose 1986) – opposition or reluctance from middle managers, workers and unions. The earlier adoption of computerised production, such as NC, had been piecemeal, and constrained partly by craft unionism. However, the scale of FMS investments has often required senior managers' involvement, as union job controls have weakened. The diminution of workplace unionism since 1979 should not be over-estimated. A wide union presence with important residual powers and indirect influence persists (Millward et al. 1992, ch. 9). Yet management have generally had more freedom to reorganise the shopfloor. Has this freedom been used to apply flexible production technology innovatively, or just to extend and intensify more rigid, conventional paradigms?

Chapter 3 examined the influence of the Taylorist–Fordist image of North American efficiency on state and enterprise industrial policy in the post-war period. This paradigm has only recently been challenged by a belief that Japanese success was based on alternative approaches. In general, the inferiority complex of 'catching up' meant adopting US standards of product standardisation, economies of scale and labour-saving production technology. These standards are inherently problematic in small-batch production firms and sectors, yet they have nevertheless become yardsticks of efficiency for managers in such enterprises. FMS in Britain spread largely through government investment schemes. At a technical rather than a strategic level these schemes tended to emphasise quantitative gains of the Fordist kind. The first part of this chapter, therefore, examines the background and detail of the relevant policies, and the second part their relationship to the choice and operation of a sample of recent FMSs.

Account must also be taken of Britain's variegated industrial structure. The oldest industrial nation consists of various strata of firms and sectors some of which, especially in metalworking engineering, still have traces of localised and craft attitudes from the late nineteenth and early twentieth century. These coexist with, or are overlaid by, the 'rationalisation' arrangements in imitation of US practice, or by takeovers of the middle of this century (cf. Nichols 1986, pp. 149–151). On the other hand – to continue the geological analogy – in the uppermost layers are more innovative firms and sub-sectors, such as aerospace. They compete in world – as opposed to national and regional – markets, and their managers participate in cross-national joint ventures, international industry forums and professional bodies.[1] Graduate engineers and managers trained in, and open to, contemporary techniques of product development, marketing and organisational change run these firms. The third part of this chapter explores the ways that a firm's place in this complex industrial structure influences FMS adoption and use.

A final factor concerns the craft-style unionism and informal shopfloor administration which managed to absorb Taylorism. Academic observers of British industrial restructuring dispute whether a combination of Thatcherite industrial relations legislation, free-market economics and a revanchist spirit of managerial control has broken such institutions (cf. Rose 1994). How far have these forces helped new forms of social organisation of production, more conducive to FMS-type technologies? The last part of this chapter examines whether enhanced managerial control and the quest for cybernation have displaced the craft union job controls noted earlier. Are industrial relations changes working with, or against, the grain of production strategies? Are they pushing FMS work organisation in Tayloristic, Fordist or flexible workgroup directions, or just preserving traditional forms?

Promoting factory automation: the art of the state

The proto-FMS that Williamson built for the Molins company in the 1960s long remained the unique British contribution to factory automation. As chapter 4 showed, the 1960s and 1970s were a period of uncertainty and competition between rival production technologies in Britain. NC and CNC won out over the more organisationally demanding Group Technology; and the appeal of Molins' System 24 no doubt suffered from Williamson's organisation of it on Group Technology/integrated cell principles. British industry was not ready for technical and organisational change of such complexity. Nevertheless, amongst the network of professional engineering associations, industrial research, and industrial policy bodies – linked to the rise of economic corporatism – engineers and managers were looking for new technological solutions to the perceived problem of Britain's industrial inefficiency.

Reports of the first FMSs in the USA and Japan reached British industry in the early 1970s. But there had been no interest in System 24, so British machinery builders had no experience, even of prototypes, of such systems. The Department of Trade and Industry's Mechanical Engineering and Machine Tools Requirements Board, a mixed governmental and industrial body, commissioned in 1976 a National Engineering Laboratory report 'on the UK and international activity in small batch production . . . its relevance . . . in the next decade' and 'to draw up a skeleton plan for the evolution . . . of a completely automated system for small batch manufacture'. This ASP (Automated Small-batch Production) Committee (National Engineering Laboratory 1978) set the benchmarks for later government policies on small-batch automation. The Committee, and its Working Party report, sometimes referred to 'flexible manufacturing systems', but conventional cost-saving production economics dominate their analysis. The Production and Engineering Research Association's economic analysis section of the ASP Working Party Report justifies an ASP system in terms of:

- reductions in direct labour and associated overheads;
- increases in productivity 'per spindle';
- reduction of work in progress and inventory;
- reduction of lead times.

There was little or no discussion of the capabilities of innovative production, or qualitative gains in the flexible response potential of FMS. The ASP Committee chairman confirmed these signs of a Fordist mentality. He told the British *Production Engineer* magazine that small-batch automation meant British firms: 'using very high technology to manufacture low technology products cheaper than the rest of the world' (*Production Engineer* 1981). Significantly the PERA contribution played down the idea of using the proposed systems for rapid shifts between very small batches or one-off designs. The extra cost of creating new programs would require a batch size of at least ten units to be financially justifiable (ASP, p. 8). The Report spoke of the need for all personnel in an ASP system to be 'flexible in terms of attitude, and where possible specific skills', but separation of jobs into 'program development, writing and editing' and 'shopfloor functions' indicated a Taylorist view of work roles. Strangely, in the light of the already available US experience, the Report conducted a *de novo* analysis of software and hardware problems and requirements, with little or no reference to overseas development. Nevertheless it ignored authentic vertical flexibility in task repertoires by recommending, instead, a typical US form of occupational specialisation on the shopfloor:

a loader, mechanical maintenance, electrical maintenance, inspection and supervision, and tool setting jobs (ASP, p. 7).

Following the ASP Report in 1977 the Labour Minister for Trade and Industry, Tony Benn, authorised a DTI investment scheme, in which the Department would pay a proportion of the costs of studying and developing a home-grown FMS-type complex. Only one of the machine-tool firms to whom this scheme was outlined responded with a concrete proposal. This was the 600 Group, which began the construction of a lathe-based FMS in 1979, the year in which the Labour administration fell. Under the Conservatives the philosophy of technology policies for industry changed from direct intervention to a doctrine emphasising that 'it is for industry to take the initiative' and government's role to 'create the right climate for the use of Advanced Manufacturing Technology' (ACARD 1983, p. 2). In practice, however, the Conservatives' successor to ASP, the FMS Support Scheme, was a much more successful interventionist act than Tony Benn's initiative. By 1984, when its funding expired, the FSS had helped set up thirty-three FMSs; several of whose managements admitted that their companies would not have been able to accept the financial costs without the government subsidies (Scott 1986, p. 274).

The continuity in the Tory and Labour policies towards technological rationalisation and investment may be partly due to the 'new technology' panic of the late 1970s and early 1980s. The potential of micro-electronic circuitry to cheapen and expand the applications of a variety of information and communication technologies exploded in the British media around 1978. Particularly influential was a BBC television programme, 'The Chips Are Down', whose mixture of futurist prophecy and warnings of national backwardness set the tone for media debates, reports and policy orientations for several years. In this atmosphere even a *laissez-faire* Tory government had to be seen to be taking an active role in promoting take-up and exploitation of the new miracle technology. This factor apart, the production technology schemes also conformed to longer-standing aspects of British industrial policy.

Since at least 1918, British policy makers had assumed manufacturing enterprise in Britain was too fragmented compared to rival economies. Early attempts at greater concentration amounted to little more than voluntary co-operation within sectors (Hall 1986) and, in the 1960s, tacit encouragement of mergers and takeovers (Graham 1972; Young 1974). Nevertheless the post-war policy ethos was consistent with Marshall Plan standardisation criteria. It assumed industrial competitiveness needed large-scale production operations run by giant enterprises (Hirst 1989, p. 284). Indeed, another Fordist heritage of the Marshall Plan years complemented such axioms: the idea of technology for large-scale product specialisation.

This idea is reflected in orientations to technological support amongst both policy makers and many practitioners. It was agreed in the House of Commons' Industry and Trade Committe that issues such as quality and reliability were less important to subscribers to the FMS Support Scheme than a balance-sheet economic justification (UK Government 1983, p. 16). The Institution of Mechanical Engineers recommended that the 600 Group FMS, and another run by the KTM company, could 'manufacture components for the whole of the UK machine tool industry' (UK Government, 1983 p. 54). Sir Jack Wellings, chairman of the 600 Group, described FMS as suitable for both large- and small-batch production, but his company's strategy as making standard products (UK Government, 1983 pp. 153, 158). But the flexibility viewpoint was emerging. The DTI defined an FMS in its guidelines for grant applicants as including 'production of a wide range of products in small numbers' (DTI 1984). User firms told Capes (1985) that these divided into applications for specialised production of a small range of parts, and those aiming to make single items allowing immediate supplies of sets of components for final products and sub-assemblies. However, such aspirations have not been widely realised, as the following evidence shows.

All the FMS Support Scheme funds were spent between 1982 and 1984, equipping thirty-three firms with FMS. The subsequent, and broader Advanced Manufacturing Technology scheme subsidised a few more installations (Sim 1984). However, the very success of these schemes in prompting a rapid adoption of FMS, raises questions about the suitability and purposes of the FMS in the adopting firms. A major reason for adoption was the availability of considerable state financial support. Thus either the investment procedures of British firms inhibit upgrading of production technology (cf. also Pullin 1987), or FMS's specific qualities were not the only improvement; or, alternatively, some combination of both these factors is influential. Case study evidence supports the latter proposition.

The rest of this chapter is a detailed analysis of data collected by Peter Scott's survey of fifteen of Britain's first FMS firms.[2] Several of these would not have been adopted without the government grants. The ASP ethos also persisted; with many, perhaps most, using the systems either for various kinds of cost reduction, or to get familiarity with a new technology; rather than for *product flexibility*. The next section explores the importance of these strategies. Notice for now that level of product flexibility – capacity to vary programs and machine functions to make different types of product – only mattered in five of these fifteen firms. Another six had not adopted, though they had considered, FMS or had corrected production requirements, because of financial controls and/or feasible alternatives to FMS. On failing to qualify for the state schemes two such firms dropped FMS plans for more modest arrangements. A neo-Fordist view of FMS led the others to think that they could not easily make small batches of varied parts. Other deterrents were lack of appropriate expertise, and related fears about the complexity of FMS planning and programming. All these firms had mixtures of CNC, DNC, and automatic part and tool handling, but not the level of integration of constituent systems which defines an FMS, at least on the criteria in the state support schemes.

Investment criteria and operating constraints

How are the British FMSs being used? Here it is useful to ask the same kinds of question that Jaikumar put to his US and Japanese FMS users. How well were managers exploiting the product flexibility and automaticity potential of their systems? Were they closer to Japanese or US styles and levels of utilisation; and amongst those with sub-optimal levels, why was this? To answer these questions requires an understanding firstly of managers' aims in commissioning the systems; and then, secondly, of the range of parts being made on each FMS. Relatedly, what were the reasons for greater or lesser product flexibility and the influence of the designated objectives?

Quantitative and qualitative aims

As the previous chapters indicated, the objectives of FMS investment may be divided into 'qualitative' and 'quantitative' aims. *Quantitative* objectives are principally those captured by accounting measures. They include all kinds of direct cost savings, such as: reduced inventory and work-in-progress (WIP) – which are reckoned as incurring costs in the form of interest charges for working capital – floorspace, and reduced labour. The indirect cost savings are: lower overhead costs – such as floorspace, energy and fuel, 'backroom' labour, etc. – as well as more continuous use of machinery, and hence faster repayment of the equipment's capital cost. *Qualitative* aims may lead to quantitative savings, but accounting practice cannot easily measure their effects. These, in turn, fall into efficiency and capability gains; including greater hierarchical control over labour and the production process. A qualitative gain in capability, on the other hand,

Table 9.1. *Objectives of FMS investment by British firms*

Quantitative Number of citations		Qualitative Number of citations	
Reduce WIP or inventory:	4	Greater control of labour & production:	3
Reduce direct operating costs (e.g. labour):	4	Reduced lead times:	2
Make production of small batches more economic:	3	Higher quality:	2
Higher machine utilisation:	2	Gain experience/show capability to clients:	1
Greater economies of scale	1		
Total quantitative:	14	Total qualitative:	8

Source: adapted from Scott 1987.

would signify a reduced 'lead time' between the design and planning stages, and final production. Qualitative capability benefits also include: higher quality in terms of the product's finish, precision and reliability, and – more importantly – a capacity for a more varied range of components or products without excessive extra planning and controls.

Most quantitative and qualitative efficiency aims are typical of, but not exclusive to, Fordist uses of advanced technology. Qualitative capability gains, on the other hand, correspond to elements of post-Fordist paradigms such as flexible specialisation. Table 9.1 classifies the various objectives cited by the FMS managers in Scott's survey, into quantitative and qualitative categories. Each firm gave more than one reason for adoption so the number of responses exceeds the fifteen firms questioned. Two inferences can be made from these responses. Firstly, quantitative aims are almost double those of qualitative gains; indicating a marked emphasis on readily calculable financial advantage. Secondly, amongst the qualitative aims, the most cited conventional Taylorist principles of enhanced control over labour, or the detailed final stages of production. The other qualitative aims are either practical improvements, product quality and gains in experience, or are ambiguous on efficiency or capability. For example, shortened lead times may mean either an aim of product variety in small batches, to enhance market responsiveness, or a much less ambitious desire: that the mere eventuality of a new design or design modification, of whatever frequency, should not entail delays in production and the associated costs of idle plant and equipment. A different classification of these and other FMS firms' investment motives identifies four aimed at 'flexible manufacturing', four at just-in-time gains, and thirteen at volume production criteria (Lee 1996, p. 106).

Consequences of quantitative priorities

Admittedly, most of the British FMSs are, or were when studied, at early stages of use and development. Changes in operating priorities could lead to something approaching flexible specialisation; as at Alpha in the USA. However, it is still accurate to conclude that the strategic aims and operating criteria of almost all the British cases constitute severe constraints to pursuit of such post-Fordist flexibility. The problems and shifts in operating practice reported by the FMS firms in Scott's study reinforce this diagnosis.

> Senior management pressures for immediate production and high levels of machine utilisation were observed at most plants and were a particular source of complaint at plants G, K, M, N and R. The grievance of plant engineers and operational staff was that they were thus denied adequate time to develop the control and flexibility capabilities of the FMSs concerned to accept broader product spectrums. (Scott 1986, p. 154)

It should be noted that these specific plants were all divisions of British-based firms that had recently experienced cost-driven rationalisations in labour and operations – a point that will be amplified in the next section. Technical problems had also reduced the flexibility aspirations of some of the firms: two systems had originally been aimed at a radical shift away from production of batches of the same components to the simultaneous manufacture of single parts that would be ready for immediate assembly into sets. In one of the latter cases there were technical control problems and, in the other, the time taken for resetting of the machines was judged to be too costly. In at least two other firms the tooling needed to produce the intended range of parts proved to be beyond the capacity of the automatic mechanisms.

The results of the predominance of quantitative cost-minimising standards in FMS adoption show up in the range of parts actually produced. Table 9.2 shows the number of parts, or 'part-families' – of similar shapes and dimensions – made in each of Scott's FMSs in 1987. The framework of financial controls over manufacturing operations seems to be a crucial factor behind their restricted flexibility. The original cost-minimising focus of state factory automation policies and recommendations figured in and contributed to this normative paradigm. Within firms operations managers and production engineers have sometimes sought to exploit the systems' potential for varied product ranges. However, pressures from senior managers and financial controls to get early utilisation and to minimise 'down-time', have limited the time and resources available to experiment and develop new programs and capabilities.[3] British manufacturing operations depend upon the form and duration of their financial ownership in larger company structures: often unstable links which restrict technological planning and policies (Small 1992).

Use of FMS and type of firm

Two types of contemporary organisation and structural position define most British metalworking operations. One type is product-based divisions of very large firms with international markets or corporate organisation. The other type is British-based firms focused mainly – at least in the recent past – on domestic, or ex-imperial, markets. In the former category are many aerospace and automotive firms, plus some subsidiaries of US corporations. Whereas most British-based makers of traditional metalworking products – industrial components, machine-tools, non-automotive engines, etc. – remain in product markets described by Freeman as of middle-value and middling sophistication (Freeman 1985). Often these firms remain quasi-autonomous in terms of their operations; integrated only by financial controls within 'federal firms' (Loveridge 1981) – larger conglomerate enterprises. Significantly for this Anglo-centric category the heartland of traditional metalworking enterprise in the West Midlands region had only one genuine FMS at the end of the 1980s (Mahadeva 1986).

Differences in general investment and production strategies correspond to these different types of firm (Jones 1988); but it is not a hermetic distinction. Some traditionalist, domestically oriented firms depend on overseas markets, and some of the internationally focused firms retain strong elements of localism both amongst management, and in some shopfloor practices. However, the systems of FMS firms in the former category are less likely to be part of a general strategy of computer integration and product and process upgrading – a more likely feature of internationally oriented firms. In such firms FMS is neither a tentative experiment with more complex equipment, nor solely a solution to a discrete production problem or new product. For firms like the aerospace companies in Scott's sample, FMS is one part of a broader plan for complete cybernation; integrating design, planning, logistics and production.[4] The fourth section of this chapter shows that such strategic technology approaches may also include plans to simplify work roles to minimise shopfloor inputs. By contrast, even if the localist–traditionalist firms sought such aims, concerted reorganisation of tasks and jobs was rare.

Characteristics of more and least versatile FMSs

Neither FMS objectives nor product range corresponded directly to type of firm, in relation to the flexibility criteria in the previous section. However, structural position is relevant to the style and strategy of operation which influences the degree of product flexibility. In descending order the firms with the widest product ranges were:

- a large, divisionalised, British aerospace firm;
- the British division of a US shoe-machinery firm;
- a British 'federal-firm' machine tool-manufacturer;
- a British industrial conglomerate's machine-tool subsidiary;
- a large producer of diesel engines, recently bought by another such conglomerate.

Table 9.2. *Range of parts made on British FMSs*

Company	Range of product parts	Comments
A	7	15 planned
B	1,000	Small similar parts
C	190	
D	10	
E	5 part families	made for sets
F	13	
G	20 parts or part families	
H	15 part families	
J	25 part families	
K	6	
L	14	
M	15 part families	
N	20–25 parts	
P	8	'sets' abandoned
R	107	

The first three – B, C and R in Table 9.2 – of these wider-variety producers, stand out. The FMSs of the shoe machinery maker and the machine-tool specialist are set up to make over 100 parts, and the aerospace firm over 1,000.

The FMS is just part of a broader scheme for CIM. Several government grants, plus the high levels of advanced engineering expertise maintained by, mainly military, firms in aerospace provided the resources in depth for the extensive programming and planning needed to develop the wide range of plans and software. Government military contracts are the main source of the longer-term capital for R&D permitting such advanced technology strategies in aerospace firms like B (cf. Jones 1985a; Robson, Taylor Hitec in UK Government 1983, p. 189).

The FMSs with the lowest degree of product variety, less than twenty parts, tended to be mainly in domestically oriented, more traditionally organised, firms. Reasons for their limited versatility relate to the character of the firms in which they operate. Table 9.3 shows their sector and organisation, and the range of products for which the FMS produced parts.

Table 9.3. *Type of firm and sector for FMSs with lowest product range*

Firm	Type of organisation	Sector	FMS Products
A	Division of British	mining machinery	7
D	US subsidiary	construction vehicles	10
E	US subsidiary	pumps	5
F	US subsidiary	machine-tools	13
H	Division of British	mining equipment	15
K	British group	automotive components	6
L	Division of British*	public transport vehicles	14
M	British group	diesel engines	15
P	British group	aerospace contracting	8

* Since taken over by a continental vehicles firm.

The machine-tool *firm R* had FMS partly as an extensive experiment in a production and legal entity separated from the rest of the firm. Because its capability hinged on a special, perhaps unique, scheme for work roles this specific case is detailed in the next section. Closer inspection of the other two high-variety FMSs shows that their product flexibility, does not entail a post-Fordist manufacturing paradigm. *Company C* had been alternately starved of investment capital by its American parent, and ordered to rationalise plant and labour. Employment fell from 4,000 to 1,700 between 1981 and 1984. Despite the FMS Support Scheme grant, the subsidiary firm's managers had to struggle with the corporate management to get extra funds for the host computer to run the machining system. Former adviser roles on the ASP programme by two of the local managers partly explains the commitment to FMS. Moreover, in line with the ASP philosophy they see the FMS as part of a process of cutting unit costs and reducing work-in-progress. In similar neo-Fordist vein a Group Technology rationalisation of the range of components preceded the FMS.

Indeed, many of the 190 parts programmed for the FMS fall into a smaller number of part-families. Moreover, the machining processes are all similar; with a six-inch cube configuration for most individual components, short time-cycles, and only a few drilling and tapping – as opposed to the more complex milling – operations. The fact that all parts are for a new line of a shoe-making machines assisted GT rationalisation and FMS planning. Thus it is even harder than it would have been to tell whether the FMS is being used to achieve a higher level of flexibility than normal practice. Yet the cost-cutting aims, the ASP connection, and the GT scheme, all qualify an inference that the 190 part range of firm C's FMS means flexibility in the sense of versatile manufacture of a wide product range. However, the discussion in the fourth section explains in more detail how, in this firm, high product variety is a smokescreen as far as innovative tasks in work roles is concerned.

At *company B*, the other high-variety FMS of an aerospace firm's military aircraft division, a similar, perhaps more intensive, tale of computerisation as rationalisation emerges. Again, the product range of 1,000 is deceptive: these are all small parts in the size range of 600 + 300 + 120 mm for one specific fighter aircraft. However, the fact that an average batch size as low as four is made concurrently does show a high degree of complexity in programming and control of the system. But the firm is similar to firm C in having ASP expertise 'in-house'.

These firms fall into three sub-types of company strategy and structure. Firms D, E and F are British divisions of major American corporations; with automation strategies partly reflecting the typical Fordist postulates of American manufacturing. Another four – A, H, K and L – are single-product divisions of British firms selling mainly to the British market. Firms M and P operate in world markets, but structural factors substantially constrain their strategic perspective. M is part of a typical British federal firm.

In *firm P* strategic development of the production technologies is inhibited by the spread of production over several sites; and attendant pressures for cost rationalisations. Several plants, including the one which originally contained the FMS, have been closed in recent years. Firm P benefits from the indirect subsidy of its engineering staff through contracting for government defence contracts, and from its involvement in networks of cross-national co-operation with overseas aerospace firms. But it lacks the broad spread of markets and products of company B, the computer-integrating aerospace firm. The FMS is dedicated to a small subsidiary contract for parts for a Royal Air Force fighter plane.

Firm M is a classic British federal firm. It is a subsidiary of a British parent firm owning other diesel engine firms, but lacking overall integration through a divisionalised structure. The FMS is small, relatively cheap and was part-financed by a DTI grant. Senior managers claim to be moving towards higher levels of computer-integration of production functions; but progress has been limited. As eleven of the fifteen parts made on the

FMS are for similar types of flywheel, the range is somewhat exaggerated. In all three American subsidiaries the approach to FMS was significantly influenced by their US parents.

Amongst these three, FMSs were installed in *firm D* as part of a global strategy to cut lead times and make smaller batches. The strategy was experimental in the sense that methodologies of large-batch production had previously been paramount. This firm was thus departing from a strict Fordist approach. In the event broader business pressures intervened leading to closure of the UK plant with recriminations and bitterness. *Firm E* was the only European plant of the industrial hydraulics division making gear pumps, control valves and cylinders. Cost excesses also dominated its production organisation and investment decisions. The choice was either to cut under-utilised capacity or to boost sales in order to fill this gap. The FMS was bought to enhance the capacity to produce a new range of pumps more economically by reducing work-in-progress, labour costs and inventory. The parent's push for improvements initiated the new product and FMS investment, but FMS methods and organisation were not borrowed directly from the USA.

Firm F, the last American case, had direct corporate influence on FMS adoption, but home-grown methods and organisation. The US parent pioneered manufacture of automated machine-tools and systems, promoting a general computerisation of design and planning and then the FMS, engineered by its UK staff, allegedly to show the capability of such systems to British customers. A line of flexible manufacturing cells, based on elements of its own FMS was promoted at the 1985 British machine-tool exhibition. Yet macro-economic forces again helped modify the FMS plan. A sharp decline in sterling's value against the dollar during 1984 and 1985, combined with other developments to reverse an earlier policy to pursue long runs and economies of scale by concentrating prismatic parts production in the USA. The British FMS was pressed into service to help with extra capacity, and the cost economies. But added to this approach was the aim of raising the FMS's capability to make various items simultaneously. In line with the craft controls culture in the locality and shopfloor, FMS operators' roles extended to scheduling.

Do similarities in strategy or structure amongst the four subsidiaries of the British conglomerates and industrial groups account for their low flexibility in product ranges? *Plant L* has a relatively high number of part types on its FMS but these are all components for one particular bus gear-box; the site's workforce being split equally between one plant for the axles and gear boxes, and the other making the chassis. The aims of unattended running and high machine utilisation confirm what might be expected of an FMS in the automotive sector: essentially this is for Fordist-type volume production with the FMS allowing absorption of the minor variations in product without disrupting continuity of production. The fact that the FMS purchase was only financially acceptable because it was subsidised by a DTI grant confirms the top management's limited interest in flexible technology.

Firm K's story is similar. A small sub-contractor to the motor car industry, it apparently gained from more out-sourcing of components by car firms; growing from five employees in 1976 to 510 in 1985. Despite a small scale of operations, with employees spread over four sites, production volumes were at car industry levels: minimum batches of 3,000 parts and averages of 10,000 were frequent. Receipt of half of its orders from one multinational car firm heavily influenced production plans. Top management only agreed the production managers' FMS plan because the DTI grant gave one-third of the cost; the investment would not otherwise have met the firm's payback limit. Moreover, the aim of the FMS – to maintain volumes and accuracy for new orders through speedier and surer controls – matched Fordist norms. Despite its small size, as a dependent sub-contractor to automotive firms, firm K was a Fordist outpost at the edge of the batch production sphere.

Firms A and H are similar in their structure and product market. Both supply mining equipment, the state-owned monopoly, British Coal, was their principal customer. 60 per cent of firm H's output went to British Coal. The formal structure of these firms resembles North American corporations' organisation into specialised product divisions. Firm A makes nothing but mining equipment, and the division that purchased the FMS only coal-cutting machinery. The latter project – one of the largest in the UK – was a major management exercise, planned over three years before implementation at a cost of over £7 million, *without* a DTI grant. By contrast, at firm H – part of a large group with aerospace and computing businesses – only DTI support secured corporate financial approval for their £3.2 million system. This business is dedicated to making the necessary valves for the mining equipment operations, and its engineering managers seem to have had less influence over operations strategy than at A. This FMS makes fifteen different parts, but nine of these are in very similar part-families. The priority given to maximising machine 'up-time' reflects this direct cost perspective. Use of automatic probes to test for accuracy was rejected, as the tests would interrupt machining.

At *firm A*, machining delays also discourage complete use of automatic probing, but only lack of tool storage reduced the planned range of produced parts from a planned fifteen, to an actual seven. Management aimed to add one more part to the FMS each month, and a third of the firm's development engineers worked mainly on the FMS; assisted by students seconded from a nearby technical college. Moreover, the small product range partly results from the physical size of the parts, which can weigh up to two-and-a-half tons and spend nearly four hours on the FMS. Despite its limited product range firm A is more similar to the first, computer-integrating, group of companies in the degree of 'top-down' commitment to comprehensive computerisation for one specific model. Like firm H, the market may constrain its desire for some genuine product flexibility: its main customer was British Coal. Yet future extension of the FMS capability may bring it to the threshold of flexible specialisation: coal-cutting machines need customising to varying underground conditions and coal faces.

Thus, apart from firm A, the sub-category of British-owned and domestic-market oriented firms had a narrow product range arising from product specialisation for major customers and the dominance of financial controls. This approach minimised managerial interest and commitment to using FMS for product variation. It sustained a dominant paradigm of Fordist efficiencies, rather than a post-Fordist flexible specialisation. Similar structural features – financial constrictions and specialisation of plant to particular products and customers – typify the whole group of low-variety FMS users and severely constrained technological advance in three of the four firms owned by British groups. Machine utilisation frequently overrode other possible gains. In the sub-category of US subsidiaries the parent firms' strategic concern with various cost efficiencies shaped most FMS adoption and use. At the margins of this entire low-variety group were the coal-cutting enterprise, firm A, and the partly exceptional US machine-tool subsidiary, firm F. This aspired to a broader product range, plus some organisational potential from its programming resources, or from the flexible work organisation between shopfloor and engineering departments.

FMS, labour and work organisation

Chapter 5 showed that trade union policy, traditions of craft regulation of shopfloor work, and equivocal managerial policies helped British operators retain a role in NC machining; even on the 'frontier of control' between programming and operating. Operators' part-programming role should not be exaggerated. Often it is limited to operators 'proving out' the first execution of a new or revised part-program, or making small modifications – 'editing' the program; often by means of limited-choice software routines such as Manual Data Input. However, compared to the USA, larger numbers of British machinists had an input to the programming of CNC. Whittaker's comparison suggested British managers accepted operator involvement in programming more than the Japanese. Has this role continued or diminished with FMS?

The options for work roles

FMS provides more radical options for reduction of the role of shopfloor staff. It downloads part-programs automatically from a central computer to the CNC controls on individual machines. Computer-controlled schedules also direct the movement of tools and workpieces into, between and out of the machining system. Interlinking of the various operations at different machines, their interdependence in an overall plan for different parts and products, means that *ad hoc* decisions and modifications at particular work stations, or for specific parts, may all threaten its overall running. On the other hand, as Zuboff points out, new skilled inputs are created. Staff can get 'real-time' information on operating problems and progress directly from the central computer system. In FMS machining some manual tasks, inherited from conventional machine-tool usage in the NC/CNC phase, are eliminated entirely. Tool setting and workpiece transfer are automated.

However, the North American survey showed how specific conditions could vary the capture and transfer of these tasks away from the shopfloor. A Taylorist transfer of conception from production workers might allow certain organisational efficiencies. But, as the Alpha case shows, the irreducible wedge of experiential skills can be expanded to: monitoring the system, some part-programming, and computer scheduling by shopfloor workers. Britain's more rapid introduction of FMSs, in the early 1980s, coincided with a national change in the balance of power between managers and unionised workers. The longer-term implications of this shift of shopfloor power is still unclear (see Edwards et al. 1992). But in strongholds of craft control, like metalworking, firms adopting FMS may now possess more power to redesign work roles than in the earlier NC/CNC phase. How are they using that opportunity?

The evidence from earlier chapters suggest four main possible types: rigid Taylorist, craft-Taylorist, neo-Taylorist and polyvalent-autonomy. A *rigid Taylorist* scheme completes the Taylorisation of work roles, stalled in Britain's post-war and economic growth periods by weak managerial strategies and trade union strengths. In an FMS it would limit workers' roles to a few routine and basic, remedial mechanical tasks; perhaps by narrow occupational sub-divisions. FMS software also has the potential to increase and centralise managers' control by monitoring and reporting on execution of workers' tasks. The hybrid *craft-Taylorist* model, which accommodated scientific management's task specialisation goals to British craft norms of skill demarcation and self-planning of tasks, could be extended in FMS. As chapter 5 showed, this role typically means various unstable compromises over skilled workers' 'vertical' powers to initiate, modify or test programs.

The other two work role types mean extending production workers' range of tasks and decision-making powers either horizontally, or horizontally *and* vertically. In a *neo-Taylorist* option automation of many manual tasks – tool setting, program selection and workpiece transfer – could allow a wider range of mechanical tasks, at a similar 'horizontal' level of competence for some workers. Checking tooling and fixturing, inspecting machined parts, remedial and preventive maintenance are typical additions. Each worker in an FMS may have some such 'horizontal polyvalence' with variable amounts of task sharing.

However, ensuring rapid corrections and refinement of programmed operations needs *independent polyvalence* for some operators (Jones 1983). Like a neo-craft model this case is based on conventional craft engineering skills, but it extends to operators 'trouble shooting' system-level and software-related problems. *Collective polyvalence* signifies an autonomous team of such workers, similar to Alpha. Scott's firms exhibited at least one instance of all of these types of work organisation. Despite a general Taylorist bias most firms lacked either of the polar extremes. *Ad hoc* and pragmatic versions of the first three models predominated. A diminished union influence still limited systematic attempts at rigid Taylorism or neo-Taylorism.

Rigid Taylorism and neo-Taylorism

Most managers saw the introduction of FMS as a centralisation of managerial control and a break with the craft controls persisting elsewhere in the plant. Some wanted FMS particularly to eliminate the elements of operator involvement in CNC programming in other departments. In practice, however, such centralised control sentiments were only partially translated into a fully articulated system of Taylorist or neo-Taylorist job design. In one full-blown attempt at firm C, the US shoe-machinery subsidiary, the supervisor had no powers to change operating or schedule programs, which were the preserve of programming engineers, to whom FMS operators could only send written recommendations on part-programs. Partly to compensate for this restricted role – to 'keep them busy' –

operators were given almost all inspection tasks – a change also aimed at cutting indirect grades of labour.

At plant H, making valves for the British mining equipment firm, the drive for central control was aimed also at the programming technicians. Their workgroup controls over part-programming were equated with the manual craft controls over production tasks. The technicians' tasks were largely transferred to a new post of manufacturing engineer who co-ordinated all tooling, part-program completion and amendments, as well as choice of methods. Work roles for production workers in this firm became neo-Taylorist: the FMS operators must complete and check machine settings and inspection. This broke with the tradition of union-regulated 'horizontal' demarcations in the rest of the plant, with one skilled operator to one CNC machine-tool.

Previous conflicts over control of programming continued to influence FMS work organisation at other firms: the major and minor aerospace firms, B and P, and the US earthmoving vehicle subsidiary. At this, firm D, operators were 'horizontally' flexible between mechanical monitoring and adjustment tasks. At firm B managers had resisted operator involvement in CNC programming, alleging risks to otherwise overriding quality or accuracy standards. The complexity of FMS operations legitimated such arguments. Operators' roles were chiefly monitoring and adjusting tooling, plus first-line maintenance. On CNC the union at firm P had kept a ratio of one operator to one machine, plus program editing tasks. However, the FMS took up all of a small satellite plant. Operators were selected for their receptiveness to horizontal task flexibility. They could not modify schedule or part-programs apart from minor tool changes, while union opposition to horizontal flexibility kept tool setting and final inspection as specialised jobs.

The survival of craft-Taylorism

Plants A and J assigned 'minor' program changes to operators if also notified to the programming staff. Otherwise these two plants, A especially, have conventional work roles. The coal-cutting machinery firm, A, employs specialised setter/ operators, semi-skilled ancillary workers for cleaning and transport; with specialists from the main plant for maintenance and inspection. This FMS lacks fully adaptive controls – one machine even has to be started manually – so the six machining centres have ten operators; a higher ratio of workers than normal for 'stand alone' machines in the main plant! At firm J, the machine-tool subsidiary of the British 'federal firm', the ratios are reversed: one machinist for each of two cells, made up of three machining centres. Once again, however, the appropriate departments do most maintenance and inspection. Yet, the FMS operators' 'vertical' responsibilities are more restricted than on CNC where machinists make some changes to part-programs.

Thus, at seven plants – C, H, B, P, D, A and J, – operators' roles had either remained restricted, or become more so, compared to other operator jobs in the plant. These restrictions arose either from managerial policies to limit or exclude operator programming tasks; or from failures to expand work roles horizontally to include maintenance and inspection. In another eight plants there was a spectrum of task combinations, arising from different traditions and exigencies. At one extreme, firm R, there was a systematic scheme for an autonomous work-group with polyvalent roles. At the other pole were the informal and pragmatic arrangements at firm E, the US hydraulics subsidiary.

Independent polyvalence

There is a difference, however, between work organisation that grants multiple skills to individual workers, chiefly through vertical extensions into programming and control tasks, and more systematic polyvalence with task sharing, on vertical and horizontal dimensions, between the whole workgroup. In the first case, 'independent polyvalence', only a single employee in the FMS might achieve greater all-round skills. In other plants, all those in the FMS might be so qualified; but in each case the worker could independently decide on changes to the part-program or work schedule software. Independent polyvalence had various causes, but mostly the custom, or proven practicality of craft roles, influenced managers.

The FMS crew at firm N was on the borderline between independent polyvalence and neo-Taylorist flexibility. Flexible exchange of the tasks of setting and adjustment of tools and workpieces was expected, as well as horizontal extension of some inspection tasks. Yet operators were also expected to edit part-programs. At firm E very low staffing contributed to wider operator responsibilities. A single operator on each shift, with white-collar 'staff' status, jointly proved out part-programs with engineers. Similar roles and status existed at firm L, the formerly British bus manufacturer. But, in addition to joint 'prove out', the four operators per shift also allocated tool setting, maintenance, inspection and loading tasks amongst themselves.

In firm F, the main British plant of the US machine-tool corporation, a team of part-programmers was 'dedicated' to the FMS, but the two operators modify programs and the system scheduling was manually executed by them. Despite the multinational corporate milieu of senior management, shopfloor organisation was strongly embedded in craft production practices and work roles of the surrounding traditional West Midlands metalworking industry. A craft paradigm also influenced firms G, the maker of seals for marine propulsion units and, paradoxically, in K – the small mass-production auto industry sub-contractor. In the latter firm management had pragmatically settled for a continuation of the operator programming that had accompanied an earlier investment in CNC machines. With batch sizes in the thousands, reprogramming was infrequent, and the new category of setter programmer on the FMS was assisted by semi-skilled workers who monitored the machinery and performed quality checks.

Firm G was distinctive in two respects. Firstly, because there had been no previous experience with NC or CNC, management had to organise from scratch the division of labour between programming and production tasks. Secondly, the initial plan for work organisation – a hierarchical specialisation with no programming tasks for operators – broke down. The absence of a programming–operating split and traditions – what Scott refers to as: 'a small tightly knit, but highly skilled workforce, to which management [traditionally, BJ] granted very considerable autonomy' (Scott 1987, p. 202) – proved significant. The original rigid distinction between a 'system manager', 'cell technician' and 'cell loader' – one each, per shift – changed to a flexible overlap, of scheduling and system management between the first two grades, and part-programming and setting between the second two. The 'cell technician' was a polyvalent setter–programmer who also had some scheduling duties. Thus in all these six firms operators had – either individually or collectively – significant 'vertical' control tasks in their work roles, as well as a horizontal range greater than rigid Taylorism, and craft-Taylorism.

Collective polyvalence (autonomous workgroups)

The need to modify detailed job designs was also exhibited at the two firms, M and R, that had initially planned workgroup autonomy for their FMS crews. At M the British multi-plant maker of diesel engines the FMS was set up at a small satellite plant with one setter–operator per shift. This worker did a variety of remedial scheduling and part-programming tasks, in addition to mechanical adjustment work. However, the company subsequently closed this site, moving the FMS to a larger, nearby plant whose planning and programming staff played a wider role. The setter–operators continued to be involved in some aspects of scheduling and program modification, but now on a more informal and piecemeal basis. The dynamics behind this change were probably similar to the modification of the ambitious autonomous workgroup design attempted at firm R, the British machine-tool federal firm.

This FMS, too, was set up away from the main operations; although physically on the same site it was organised as a separate enterprise. After a lengthy demonstration period the FMS was relocated in the main part of the plant and adapted to manufacture mainly small turned parts for the firm's main machine-tools: gears, discs and shafts. At the initiation of the project managers sought a new form of work organisation to match the needs and processes of an FMS. Arrangements in various countries were studied in the preliminary fact-finding mission phase. Eventually, the management team decided on a small three-person crew with completely interchangeable tasks. Three operators and the system supervisor shared all other duties, including modifications to scheduling and part-programming. The system manager – with overall responsibility for the FMS plus initial programming of the work schedules – was appointed late into the project; partly because the operators had spontaneously begun to concentrate on specific tasks, which they regarded as 'theirs'. Managers hoped the system supervisor would ensure a return to the original undifferentiated task sharing; but this plan was thwarted by broader forces when the FMS was set up in the main plant.

Each of the FMS crew had initially been put on to staff status, partly to reflect their new range of responsibilities. In British firms white-collar 'staff' employees have traditionally had better terms and conditions of employment, but not necessarily pay, than manual 'works' or direct employees. But staff status created problems when the FMS was integrated into the plant's main operations. Personnel managers were negotiating a new bonus payment scheme with unions at the time of the reorganisation. The production engineer in charge of the FMS and the crew proposed a new category of 'cell technician' to cover the polyvalent role of the FMS crew. Similarly to US cases, personnel management resisted this proposal, fearing more complex renegotiations of manual workers' pay. So the FMS group were offered the choice between 'works' and 'staff' gradings. The worker who had begun to assume more programming tasks opted for staff status. The other two – with an eye to the higher payments available under bonus schemes – chose the 'works' category. After this decision the nascent task specialisms became sharper. Now the two 'works' employees concentrate mainly on manual-mechanical tasks, and the 'staff' employee mainly on programming.[5]

Strategy, structure and work organisation

What societal factors have influenced the organisation of British FMSs as compared to those in Japan and the USA? Two main features of the British case stand out. Firstly, the category of skilled, normally craft-trained, production workers remains unaltered. Even if firms saw FMS as a means of strengthening control and securing Taylorist, or neo-Taylorist work roles, they still recruited craft-qualified workers, normally their own existing employees, into the FMS. Likewise, deliberate schemes for new, enhanced, occupational status, such as the erstwhile 'cell technicians' at firm R, were rare. Secondly, and partly explaining the first point, there was a diverse pattern of task organisation with three main forms.

1 *Rigid, quasi-craft, or neo-Taylorist* forms with clear-cut limits on operators' role in 'higher' programming and planning tasks.
2 *Independent polyvalence*, often pragmatic and craft-influenced in character, where maintenance of skills or their vertical extension, provided various degrees of programming, or planning involvement to operators.
3 *Collective polyvalence*, confined to a couple of firms where all workers' roles were significantly expanded both horizontally and vertically.

Societal influences appear more influential than micro-organisational and technical factors. Both the flexible specialisation hypothesis, and Jaikumar's similar theory about FMS use, attribute wide product ranges to craft or post-Taylorist work roles. Yet Table 9.4 shows no unambiguous correspondence, in British FMSs, between the different work patterns and operational strategy for making a broad or narrow range of products. There is some association between the product range of an FMS and the type of work organisation, but this association is not universal. Leaving out the ambiguous case of firm A – with its partly polyvalent and partly neo-Taylorist work roles – the group of five *low-variety* FMSs contained three plants with variants of Taylorism, but only one case of the independent polyvalence form. Moreover, this type of arrangement was in the car components plant with very large batches and infrequent product changes.

The British *high-variety* FMSs have an even more mixed pattern of job designs. Collective polyvalence, one independent polyvalence and one neo-Taylorist job design were all at this end of the spectrum. Amongst the '*medium*' variety producers were three cases of autonomous polyvalence, and one craft-Taylorist. Yet operating staff at the two firms, C and B, with the *very highest* range of products – 190 and 1,000 – both had neo-Taylorist organisations. So product variety in the UK is not directly linked to work roles. Indeed well-organised firms can get product flexibility from FMS technology *without* programming by skilled or craft workers. Provided, however, other expertise compensates for limited skill inputs on the shopfloor as also in some low-variety cases. As in previous sections in

Table 9.4. *Variety of products in relation to FMS work organisation*

Firm	Product variety (actual numbers)	Work organisation
E	LOW (5)	I-P / N-T
K	LOW (6)	I-P
A	LOW (7)	C-T
P	LOW (8)	R-T
D	LOW (10)	N-T
F	MEDIUM (13)	I-P
L	MEDIUM (14)	I-P
H	MEDIUM (15)	N-T
M	MEDIUM (15)	C-P
G	MEDIUM (20)	I-P
N	HIGH (20–25)	I-P
J	HIGH (25)	C-T
R	HIGH (107)	C-P
C	HIGH (190)	N-T
B	HIGH (1,000)	N-T

KEY
PRODUCT VARIETY: LOW = 1–10, MEDIUM = 11–20, HIGH = 21 + WORK ORGANISATION: R-T = rigid Taylorist; N-T = neo-Taylorist; C-T = craft Taylorist; I-P = independent polyvalence; C-P = collective polyvalence

these firms the specialist automation department and full-time manufacturing engineer part-programming, and ex-ASP committee expertise made software development easier.

This dependence on 'top-down' inputs may not be enough for reprogramming and high utilisation levels in the future. But proficiency and strategy may be more related to industrial structural factors. At both B and C the systematic CIM outlook is linked to internationally focused, divisionalised structures. As we saw in the third section above, links between the form of work organisation and type of business are much stronger than that between product versatility and type of enterprise. The firms exposed to overseas control, co-operation or competition all had neo-Taylorist work organisation. These were the British-owned firms with strong international, normally US, connections and the British subsidiaries of US corporations, B, C, D and E. The medium-sized British aerospace firm even had a more rigid Taylorist job design than the craft-Taylorism in its main plant. The remaining neo-Taylorist work organisation was in a medium-variety FMS at the divisionalised mining machinery firm's valve-making plant. Here product and production organisation had more Fordist features than most of the other British domestically focused firms, and more than the other maker of mining machinery.

Significantly, perhaps, the product range was more limited than the crude numbers of product-type suggest.[6] At the other four medium-variety FMSs, one collective polyvalence and three independent polyvalence job designs were all in domestically oriented, British-owned and organised firms; except for the machine-tool firm F. Even this British branch of the US corporation had British local management of a plant in the heart of the West Midlands metalworking region – amid strong norms and practices of craft control rooted in local labour and union traditions. Its mix of shopfloor craft norms and managerial indulgence (Whittaker 1990, p. 137, Terry and Edwards 1990, p. 162) imparted a production culture resembling British-owned counterparts. By contrast, in the higher-variety FMSs at firms B and C, neo-Taylorism matched ideas of enhanced 'top-down' CIM controls, typifying their international milieux. Yet at N and J, the other high-variety firms, British ownership, domestic orientation, and dispersed 'federal-firm' management conform to autonomous polyvalence and craft-Taylorism. The exceptional firm R was also of this type after dilution of its highly planned collective polyvalence.

Conclusion

British FMSs show neither a strict association between Taylorist organisation and Fordist standardisation, nor between variable 'flexible' product strategy and devolved or craft-style work organisation. In these FMSs, company structure and strategy relate closely to the form of work organisation and, in turn, to their flexibility – defined as variety of product types machined. Job designs could subsequently facilitate or constrain product range, but UK firms do not appear to adopt broad or specialised job designs in order to make wide or narrow ranges of products. As well as societal institutions such as occupational statuses and employment relations, the broader structures of enterprises and sectors, not just immediate managerial aims and policies (Smith, Child and Rowlinson 1991; Nichols 1986), also influence the organisation of work and production.

The heterogeneity of metalworking engineering, its mixed structure, culture and market environments contribute to enterprise and even plant particularism in terms of personnel, management, technology and market environments within sub-sectors. Multi-divisional, and often federal, firms have a varied product and managerial arrangement creating a mosaic of different planning and administrative arrangements. Lengthy chains of financial control constrain operations management. Although the distance of financial decision-makers from the production sphere may, paradoxically, allow managers there some autonomy of action in selection and organisation (Lee 1992); as in the North American FMS reported by Thomas (1994).

In the struggle to cybernate batch production the polar opposite to federally organised and locally managed firms – divisionally organised enterprises – provide another significant, if less frequent, environment. These display coherent corporate-level strategies to align technology and labour use with product and financial strategies. In British metalworking, these are more likely to be British-owned firms competing or co-operating in international markets, or British divisions of foreign, usually US, multinationals. Unlike the British domestically oriented firms, internationally focused firms are more likely to see the extensive technological upgrading of CIM, FMS and so on, as competitive options to match foreign rivals or their own foreign divisions.

The pattern of 'flexible' automation in Britain has been a 'revolution from above' in two senses. At a societal level state intervention has spread FMS and influenced its aims; while the partial 'rolling back' of union craft controls has influenced FMS organisation. It is also a top-down revolution in the corporate sphere. Enhanced managerial power has increased the scope for technological assaults on production problems. In most of the British-owned companies – and some subsidiaries of US corporations – directors and parent firms use strict financial controls for necessary investment returns to manage their constituent operations. In Scott's sample, only the state subsidies allowed FMS purchase within the resulting expected rate or time. In this sense the DTI went over the heads of the higher managers to assist production managers and engineers to justify technology investments.

Conventional modes of plant union bargaining had often shaped the first waves of the preceding NC and CNC technology; or at least its staffing and work practices. However, when most FMSs were adopted, union bargaining power was retreating from recession, government legislation, and more unilateralist management styles. Trade union influence might persist in the plant as a whole, but managers were much freer to decide workers' numbers, skills and job content. Unsurprisingly, therefore, some firms broke with previous webs of craft controls over work practices and planning by trying FMS job designs with either greater polyvalence or Taylorist/neo-Taylorist centralisation of planning and programming. However, the localised – frequently craft-encultured – plant management in most British, domestically oriented firms often restrains this managerial offensive. Evidence of planning for operations without craft-trained workers is rare. Indeed the predominance of either more craft-Taylorism, independent polyvalence, or collective polyvalence, continues a tradition of skilled production workers' influence in decision-making; despite overall decline in union regulation. This persistence suggests such forces may continue to inhibit cybernated factories. Adherence to Fordist standards, however, suggests British FMSs will probably not be used for flexible specialisation.

10 The third Italy and technological dualism

In Italy, like Japan, a dual industrial structure is crucial to the development of factory automation. Technological applications differ between small, owner-managed, firms and larger, more highly capitalised ones. Japanese dualism made possible the supply of general-purpose CNC machines to the small 'secondary sector' firms, and specialised use of FMS for cost-saving economies in the larger firms. In Italy the organisation of very small firms is more institutionalised, and their character is even more significant, for they have re-established the vitality of workshop over the Fordist logic of the factory. This technological trajectory contrasts with conventional commentaries which portray the past and future path of factory organisation (cf. Bell 1972; Kaplinsky 1984; Jaikumar 1986) as a sort of technological evolutionism: an increasing scale of automatic control and complexity; a path leading from NC machine-tools to fully computer-integrated operations, through systems such as FMS. A uniform adoption of small-batch production technology is presumed. However, we saw in chapter 4 that the earlier success of CNC machine-tools arose only out of competition between CNC, GT and integrated cells. In Italy the greater compatibility of CNC with workshop principles seemed so important that commentators such as

Sabel proclaimed craft-based flexible specialisation as the successor to the Fordist logic.

Japan showed the possibility of a dual trajectory with corresponding differences in the accompanying production paradigms. In Italy, too, larger firms and plants do indeed aim for CIM-type developments, while smaller ones adopt equipment such as CNC or CAD for special and selective uses: for use with continuing and varied amounts of traditional human skills. The associated bifurcation of production methodologies means larger computer-integrated plants are, like the US and some Japanese producers, pursuing CIM in Fordist or neo-Fordist ways. By contrast, smaller producers use new technology simply to enhance their traditional flexibility for small-batch production.

Overall, the Japanese, British and US FMSs studied do not seem likely to be leading to a post-Fordist regime such as flexible specialisation amongst larger firms. Following Sabel's original schema of possibilities in *Work and Politics* leaves us with the small-firm sector as the most likely source of flexible specialisation and any renaissance of craft production. These considerations led me to a more detailed study of the contrasts between the large and small-firm sectors in Italy's Emilia Romagna region. Emilia, between the northern Apennines and Adriatic coast, has become a key industrial region of the 'Third Italy' – distinct from the traditionally industrialised north and the semi-agrarian south. It has also become a kind of intellectual political football between opposed camps of Anglo-American academics.[1]

For the post-Marxist Left, the region has become both a symbol and a model for more democratic 'grassroots' market processes (Piore and Sabel 1984; Sabel 1982; and Zeitlin 1989). Emilia's appeal stems from its combination of localised small-scale production, fostered by communist and socialist local governments in defiance of the big business hegemony of national governments. For a diminished but articulate number of Marxian critics, on the other hand, Emilia's industrial prosperity is a corrupt chimera, based on the dubious revisionism of the Italian Communist Party (PCI) and primitive exploitation of unprotected employees and homeworkers. Is it only a temporary and localised hiatus in the longer-term advance of big business Fordism (F. Murray 1984, 1988; R. Murray 1985; Amin 1988, 1994; Matera 1985; Harrison 1989)? Now legions of foreign academics and policy experts, seeking the reality of the Emilian industrial miracle, follow where Roman armies once marched the Via Emilia.

It is easy, but not very enlightening, to see the relevance of the Emilian evidence in the terms set by these intellectual joustings. Encouraged by Piore and Sabel's over-ambitious modelling, these critiques are often unhelpful because they focus on flexible specialisation as a style of business management (Amin 1994), rather than the original definition which hinged on versatile production using interchangeable skills (Brusco and Sabel 1981). However, Emilia Romagna does have a specific significance for the themes of this book: the interplay of social and technological developments, and the current status of small-batch factories. Metalworking engineering sectors are prominent in the region with 42 per cent of manufacturing employment in 1981 (Giovannetti and Zini 1989). Like others in the region, these industries are composed principally of small firms interlarded with a few larger, high-technology enterprises. But, a point often overlooked by both proponents of flexible specialisation and its critics, the most significant element is the distinctively Italian socio-legal category of very small 'artisanal' enterprises. Thus the nature and extent of this neo-craft alternative – also the contentious point for critics of the flexible specialisation thesis – makes Emilia a crucial case.

Flexible specialisation theory equates these smaller firms with post-Fordism because, allegedly, they revive an earlier paradigm of disintegrated production by combining traditional craft skills with computerised production technologies. If true then Emilia represents, at the very least, a rival production system to the march of Fordism. From this book's perspective it means batch production gravitating back to the workshop pole, away from Fordist/neo-Fordist moves towards the cybernated factory. If technological evolutionism is correct, however, such small-firm flexible specialisation can be only a temporary deviation from eventual cybernation. Renewed competitiveness from capital-intensive flexible production, in computer-integrated larger plants, could overtake the famed flexibility of the small-firm craft model (cf. Amin 1988, p. 13). The performance of FMSs we have examined so far casts doubt on that outcome. Yet conditions which favoured small-firm flexibility in the Third Italy, in the 1970s and early 1980s, could now be assisting a new technological flexibility in larger firms. These might now compete with, and replicate small-firm flexibility. In Italian terminology, *decentramento* could be giving way to *recentramento* (Utili 1989, p. 63). The third part of this chapter examines this specific possibility.

Such a shift could also occur because of inherent weaknesses in the small-firm sector itself. Various commentators have observed deficiencies in the small-firms' technical and managerial capabilities. Sabel himself, in an early formulation of the flexible specialisation thesis, noted the risks of small entrepreneurs opting for growth via standardised products, or lapsing from innovation to routinisation in customers and markets (Sabel 1982, p. 228). Others have argued that the special embeddedness of the small Emilian firms in the human resources of family and community may now be breaking up as the younger generation of more highly educated Emiliani aim for white-collar and service-sector jobs rather than the virtues of manual skills and family enterprise (Capecchi 1989, pp. 261–3). The following account challenges such pessimism. An analysis of the social infrastructure of the artisanal firms, and their internal organisation, suggests that the dynamics of their collective growth and autonomy in previous decades still exist.

A closer look at the nature of their flexibility provides grounds for doubting that this competitive edge will degenerate or be superseded by the 'flexibility' of more highly automated larger firms. However, it will also be argued that, while a craft organisation of work is central to the original concept of flexible specialisation, this craft organisation's most vital forms derive from the institutions and social history of the artisanal enterprise. Detailed organisational and institutional factors, to which Piore and Sabel's model pays insufficient attention, account for both the successes of this alternative, in relation to high-tech Fordism, and its limitations. It is institutionalised forms of collective organisation in the artisanal firms, not just small firms in general, which give them sufficient purpose and continuity to overcome particular problems of individual small firms. This collective agency also provides resources to rival and, in some senses, exceed those of the larger firms. Moreover, the small and artisanal firms' structure is socially reinforced at the workplace level by the chances to exercise skills and responsibilities and, like Japan, eventual career paths offered to male workers. These are dynamics which the larger Italian firms, and their computerised automation schemes, have yet to develop in the same way as their Japanese counterparts.

Meanwhile, in the larger firms, FMS and related technologies are unlikely to be used to manufacture flexibly the small batches of components and products in which artisanal firms specialise. Like most US, Japanese and UK firms, FMS-type automation in Italy is more likely to be installed for larger batch and mass production, to reduce costs and shorten lead times. In addition, many Italian manufacturers tend to specialise in more customised product lines. So, unlike some of their Japanese peers, Italian capital goods producers, cannot attempt standardised production for their home markets. Indeed, evidence from larger firms in medium- and small-batch production suggests these will still combine a complex of contracting-out their special operations, with internal organisation for customised production of final products.

These requirements, plus trade unions' continued attachment to skill-enhancement, or *professionalità*, limit technological centralisation of tasks and deskilling of production work. The explication of these phenomena follows the pattern of the preceding chapters. The first section updates the Italian technological trajectory begun in chapter 5. The second section similarly continues the account of the social trajectory of employment and industrial organisation; as these factors relate to Emilia and the artisanal sector. In the third section the internal characteristics of the artisanal workshop are analysed from interviews and surveys of sub-samples of these firms. Finally, the experience of the artisanal sector is compared with first-hand evidence of the recent and prospective development of large firms pursuing flexible automation.

Technology and industrial development

Building on the springboard of expanded capital investment in the general growth phase of the 1951–71 period (King 1985, pp. 71, 75), Italy continued to improve the quality and quantity of metalworking equipment. As we saw in chapter 4 the adoption of NC, primarily CNC, machinery grew spectacularly between 1970 and 1983, such that the estimated rate of new NC was 2,000 per annum (Dina 1986, p. 9 and scheda figure 5).[2] These increases in higher technology reflected high general investments in metalworking equipment. Like Japan, these were linked to the vitality of the machine-tool production sector, and Italy's specific enterprise structure and production organisation of manufacturing. By the late 1980s Italy had the highest ratio of machine-tools per employee in the world. A figure linked by experts to the 1970s' decentralisation of manufacturing into successive tiers of smaller and smaller sub-contracting enterprises. To meet contractors' quality and reliability requirements the smaller sub-contractors were obliged to buy new machine-tools (Cazzaniga Francesetti et al. 1988, pp. 28–30). A trend which has also encouraged acquisition of NC technology and is leading, on one estimate, to a figure of 40 per cent of all new CNC being installed in firms of between 20 and 49 employees (CNR, UCIMU, Politecnico 1987).

The character of the Italian machine-tool industry has partly shaped and been shaped by these developments. By the late 1980s it was the fifth-largest national industry in the world: a position substantially linked – like the similar industries of Germany and Japan – to its share of sales, 65 per cent, of the home market. However, unlike the Japanese trend to mass (or at least large-batch) production of general-purpose machines, the Italian industry – mainly composed of about 400 small-firm suppliers and main sub-contractors – has tended to concentrate on machines for specialised uses; reflecting, once again, the segmented and specialist nature of the Italian firms which form the majority of customers. This nexus is both a source of strength – adaptability to special orders, experience in producing machinery suitable for special tasks, and access to a range of sub-contractors for special parts – and also weakness: in terms of research and development expertise, and financial resources (Cazzaniga Francesetti et al. 1988, pp. 24–5).

These structural conditions are also an important constraint on the potential for advancing technology towards CIM levels. Precise figures are notoriously scarce and unreliable, but Italy has not installed true FMS installations at the same rate as the UK, a country whose manufacturing capacity it overtook a few years ago. There was nothing comparable to the British government's FMS support schemes. The Italian government gave none of the funding from its 1982 programme for technological innovation to the metalworking sector (Momigliano 1986, cited in Baglioni 1986). However, the structural reason for this scarcity lies once more in the composition of industry and enterprise fragmentation. By the late 1980s there were an estimated eleven FMSs in metalworking engineering firms. But seven of these were in the mass to medium-batch range of producers in the auto-vehicles and motorcycle industries (*Lito* 8–9, 1987, pp. 10–11). Reports from the Italian suppliers suggested that FMS was congregating in the large firm/large batch sector of Italian industry, rather than in the small-batch firms which competed through flexible production capability (Cazzaniga Francesetti et al. 1988; B. Jones: interviews at Mandelli and Comau 1985, 1987).

Italy's FMS manufacturers themselves reflect the industrial structure's general divergence into larger, still mainly Fordist, enterprises such as Fiat and Olivetti,[3] and a sector of smaller firm networks. Olivetti companies market flexible automation systems, but Comau and Mandelli – both with major plants in Emilia Romagna (in Modena and Piacenza respectively) – have supplied most FMSs. Comau's Modena plant, like Mandelli, sub-contracts extensively. It had no computerised production, but in other respects this FIAT firm has a Fordist orientation. Most of the ten FMSs it sold up to 1987 were to big factories with large-batch and 'rigid' production schedules (Kluzer and Patarozzi 1987, pp. 7–8, 10). Comau's auto-industry focus, supplying FIAT Auto and foreign vehicle makers, means US-style production of pre-specified 'turn-key' systems.

By contrast, Mandelli tends to tailor systems to customer firms' individual requirements. This is done by drawing on a network of sub-contractors, which add another 300–400 mechanical engineering workers to its own 570 employees at the Piacenza plant. The nature of this sub-contracting network has changed in the last fifteen years. Mandelli's own role in production of machinery and machining systems has become that of a leading enterprise. Its criteria for sub-contractors have changed from capacity for standardised, large-volume, simple components and competitive prices to more specialised technological capability for preparing all aspects of components which Mandelli itself assembles and finishes. It co-ordinates, plans and provides all relevant services for clients' orders (Cazzaniga Francesetti et al. 1988, pp. 73–4).

Mandelli therefore represents a high-technology adaptation of Emilian decentralised, specialist small-firm networks, as is shown by both the organisation of its production methods, and in the kinds of specialised systems which achieve global sales. Mandelli admits that its own recently built FMS, at a new plant near Piacenza, is mainly for experimentation and development, to gain greater familiarity with potential customers' problems and solutions. Founded as an artisanal firm in the 1930s, Mandelli was itself, just thirty years ago, only another small, specialist Emilian machine-tool producer. Yet, by the late 1980s it could be described as typical of the 'enlarged artisanal model' (Rieser 1988) because of its combination of skill-intensive production organisation and the specialised character of its computerised products. Mandelli continues to evolve in size and form; but the fact that FMS can be produced by its apparent antithesis – quasi-artisanal operations – symbolises technological trajectories' potential mutability. It also highlights the importance of social developments in shaping the Third Italy's artisanal syndrome.

However, it was the internal and external relations of artisanal production that were the empirical and conceptual kernel of flexible specialisation, which Piore, Sabel et al. have since extended into a general model. These very small operations must be analysed in order to distinguish this form of flexible manufacturing from both Fordist mass manufacturing and neo-Fordist use of FMS. We will return to look at the operation of Italian FMSs as a possible alternative and threat to artisanal uses of labour and technology, after probing both the work organisation and socio-political institutions of artisanal production.

The partisan artisan: the 'social politics' of Italian industry

As described in chapter 4, Marshall Plan policies were important in reconstructing post-war Italian industry with American Fordist–Taylorist technology and methods. However, this reorientation was partial, and uneven between sectors and regions. Emilia Romagna deviated significantly from this American model and, although Sabel and other commentators have noted their influence, the decisiveness of political forces in this deviation has to be emphasised. Specially important were the national political divisions between the centre-right and the communist left which triggered and guided indigenous economic developments towards a distinctive 'Emilian model'.

Sabel's original identification of flexible specialisation in north-east Italy emphasised 'the role of ideas about the world, political conceptions in the broadest sense' and that 'the emergence of the new type of innovative economy depended on the co-incidence of various political struggles' (Sabel 1982, p. 225). But this sociologistic perspective tends to distract attention away from underlying socio-political institutions within which the ideas of groups and individuals have to be crystallised, if they are to persist and reproduce themselves. Moreover, Piore and Sabel's major statement of flexible specialisation further displaces the original insight of socio-political causality because it seeks analogous arrangements for US industrial policy (Piore and Sabel 1984).

It should be noted that in regions other than Emilia small-firm and artisanal enterprise have a significant industrial role and quite different social and political institutions. Analogous political linkages exist in some areas of Christian-Democratic (DC) government like Veneto. There, however, mutual support structures are less intensive (Trigilia 1986), and in the 'mechanical engineering' sectors on which we are focused, artisans are much less concentrated than Emilian artisans, who are Italy's third most specialised in this sector (Guagnini and Pignatti 1991, p. 51). My argument is that the interaction of socio-political identity and a corresponding institutional infrastructure ensure Emilia's specific industrial practices. This nexus of institutions and politics also explains why Emilian industry continues to differ from both flexible automation in the large firms, and the flexible specialisation Piore and Sabel allege in 'federated enterprises', 'solar firms' and 'workshop factories'.

The indispensable core of Emilian developments are not small firms in general. Rather it is a special socio-political category within the small-business sector: the artisanal enterprise. In metalworking artisanal enterprises made up almost 80 per cent of all firms in 1981 (Drudi 1986), and provided 42.7 per cent of employment in 1978 (Fillipucci 1984, p. 38).[4] Their growing centrality in the industrial character of the region arose from, and still depends partly on, the political divisions within Italian society. The economic aspirations and commercial expertise of their owners and employees are important, but secondary influences. Above all, artisans in Emilia have a distinctive economic, social and political identity which is probably unrivalled elsewhere in Italy, let alone other nations. This identity has three separate sources: Emilia's political exclusion from Fordism; a phase of political-cultural integration with socialism and communism; and the regulation of artisanal practices by legal statute.

The frequently cited alliance of small industry with the communist and socialist parties that have governed in Emilia during the post-war period, were by-products of national Christian Democrat-led (DC) government policies and the populist policies of the PCI leadership. The DC government preferred to give Marshall Plan funds and assistance to large manufacturing complexes in Piedmont and Lombardy, rather than to its potential enemies in 'red' Emilia. It chose also to run down four large armament factories in Reggio Emilia, Bologna and Imola from a total workforce of over 27,500 to 2,700 between 1945 and 1954, rather than convert them (Melossi 1977, cited in Capecchi 1989, p. 254). Many Emilian plants thus lacked the technological and dimensional basis to upgrade to Fordism. The skilled workers made redundant had to choose between self-employment or migration to the northern factories.

At the same time that national government policies were excluding Emilia from Fordism, the die of political allegiance was being cast from a speech made by the PCI leader, Palmiro Togliatti. At Reggio Emilia on 24 September 1946, Togliatti called on the communists of Emilia to recognise their shared political and economic interests with the small entrepreneurs in industry, agriculture and commerce, to preclude any reclamation of these strata by Fascist and plutocratic forces. An insight which later facilitated their identification, together with the industrial *ceti medi*, as a popular bloc against the big business supremacy in Italian national politics (Togliatti 1974, pp. 21–39). Togliatti was partly recognising a social fact. Despite a long history of urban industrial activity, stretching back to the silk industry of the sixteenth century and machine building in the modern period, the legacy of isolated sharecropping and proto-industrial cottage-industry crafts in agriculture had created in Emilia a distinct stratum of semi-agrarian, small entrepreneurs nested in kinship forms of organisation. Fascists and communists fought for the allegiance of these strata and the urban skilled workers in the inter-war period. The resistance movement of the war years then brought some of these agrarian and urban artisans under the influence of the Left in the struggle against Fascism.

Mussolini's grant of distinct legal status to the artigiani in 1936, was part of a strategy to incorporate these middle strata on a national scale. This legislation survived into the post-war period. Several statutes by centre-right governments reinforced artisans' legal distinctiveness by exemptions from some tax liabilities, record keeping and, later, collective bargaining requirements. The laws require owners to work partly on production tasks and exclude fully mechanised establishments (Lazerson n.d., p. 12). This mix of political and social developments has endowed artisans with a specific socio-legal identity: separating them from employers in larger firms and, as employers, from workers with similar skills in employment. Yet the outcome of Togliatti's injunctions in Emilia has been to bind the artigianato, as a semi-autonomous element within the communist-led social coalitions, into a complex of leftist policy-making by local government and voluntary associations.

Just after the war the Confederazione Nazionale dell' Artigianato (CNA) was formed with regional and sectoral sub-divisions such as the Federazione Nazionale Artigiani Metalmeccanici (FNAM). In the industrially decentralised Third Italy representation in all small-firm business associations is reckoned to be high. In some districts up to two-thirds of eligible entrepreneurs are members (Trigilia 1986, p. 169). The legal maximum for artisanal status is twenty-two employees, but the CNA also accepts slightly larger firms as members. The important point is that representation is higher and more meaningful than in the small business organisations of many other industrial societies, such as Britain. Moreover, the Italian associations often link the definition of political perspectives and agreements to the provision of practical services to individual members (Jones and Saren 1990).

The decisive role of the artisanal associations – particularly in Emilia that of the CNA – is translating a mass of small entrepreneurs into a collective social and political actor. Associations like the CNA voice district and sectoral concerns and aims to the commune (city) and regional governments, rather than the national level (Trigilia 1986, p. 172). So these agencies facilitate the industrial relations and local government agreements, instanced by writers such as Sabel as important for a social infrastructure of localised flexible specialisation. Thus the PCI, and often the CGIL, have favoured concessionary bargaining in collective agreements signed with artisanal organisations. Without this formal structure, based on the associations, local governments and unions could not assess the relevance of their proposals to individual artisans, for whom individualistic courses of self-interested action would have more appeal. It would also reduce the basis of vocational identity as family and neighbourhood ties diminish – as warned by Capecchi. The scope and depth of the associations' role is shown by a few examples of CNA activity.

In Emilia Romagna 54 per cent (78,084) of eligible artisanal businesses are CNA members (Rubini 1987). Cities like Modena have similar proportions of metalworking members amongst eligible firms (Lazerson n.d., p. 10). The number of branches (300), and of direct employees (24,000), indicates the regional CNA's range of activities. These include an accounting service for 43,700 enterprises, payroll for 70,000, and 110,000 individual tax returns (Rubini 1987). Special CNA arrangements with banks provide credit. CNA personnel often co-ordinate the local artisans' many consortia and co-operatives, formed for special forms of capital purchase and financing. From 1976 to 1985 Modena's 'credit co-operative' guaranteed £50 million of loans (Brusco and Righi 1985, p. 25). The Modena FNAM also negotiates directly with large firms, such as the FIAT subsidiary Italtractor, for sub-contracting artisans. The CNA has a major training role: organising courses for artisans and their employees. In firms with 10–49 employees – which includes some non-artisan firms – artisanal and similar business bodies provided training for 50 per cent of employees and 25 per cent of owners and managers (Fillipucci and Lugli 1985, pp. 202–3).[5]

For much of the post-war period, these and other CNA services have accompanied a collectivist political culture that would seem incongruous, even bizarre, in other societies:

> our Federation's first goal is to defend the interests of artisans and small business, and to demand new fiscal laws and public services; we want to obtain the amendment of laws deemed unjust and to operate for a democratic and balanced development in Italian society ... a peaceful co-existence between countries, for the realisation of peace policies and for the overcoming of all injustices. (Rubini 1987, p. 3)

In such cultural mores the spirit of Togliatti's socialist vision for the artisanal classes has interpenetrated the book-keeping, marketing initiatives, and managerial training services of practical commerce. Policies with a greater emphasis on business efficiency may now be diluting this ethos. Yet it inspired a political and social consensus for collectivist institutions and policies which promoted artisanal metalworking more than in most other regions.

Work and technology in the artisanal workshop

Emilia's successful flexible specialisation has been attributed to use of computerised equipment with craft work (Sabel 1982, pp. 220–1; Piore and Sabel 1984, pp. 205–6). This combination may have had an impact, in some industries. Yet it begs a number of questions about definitions of craft work, and computerisation in metalworking. On craft work Sabel claims in *Work and Politics* that the most advanced flexible specialisation firms fuse technicians' and owner-managers' expertise with craft knowledge. But this is not the same as craft *organisation* of work. British and US firms' promotion of machinists into NC programming jobs, shows that a synthesis with craft knowledge may combine with production methodologies other than flexible specialisation. The question is: who does what in flexible metalworking firms; and, this chapter's focus, how does work organisation in artisanal firms differ?

The role of computerised equipment is also debatable. The growth period of small firms were the 1960s and 1970s (Capecchi et al. 1978, p. 76; Sabel 1982, p. 221), especially the period up to 1975. Yet this was before the real take-off in the diffusion of NC tools in Italy. In addition, according to a variety of sample surveys, the adoption of NC and CNC in artisans' firms was low throughout the 1970s. In 1983, 18 per cent of a sample of 120 Bolognese artisans had numerically controlled machines, which represented only 1.86 per cent of all their machine-tools (ECIPAR/FNAM 1984, pp. 22, 206). Another broader survey of 830 artisans in Bologna province in 1984 found only 6.8 per cent with NC/CNC (Capecchi 1984, p. 28). Even in firms with a wider range of between 20 and 49 employees, which would therefore exclude most artisanal firms, a survey of the whole region in 1982 found only 12 per cent with CNC (Pasini 1985, p. 21). In Modena province, reputedly containing some of Emilia's most innovative and independent small firms, only 7 per cent of all investments in the period 1975–80 were of NC/CNC machines (Rieser and Franchi 1986, p. 39).[6]

None of this of course denies that use of CNC may amplify the capabilities and competitiveness of artisanal firms, nor that it has not happened since the 1970s. It simply reminds us that Piore and Sabel's characterisation of small-firm technological innovators could only have applied to a very small minority of processes and firms amongst the artisanal sector. In the heroic phase of artisanal development CNC and related technologies can only have played a minor role. If artisanal metalworking firms, and their rapid growth in such key industrial districts as Bologna and Modena, were excluded from the flexible specialisation ambit the contribution of that approach to the success of decentralised production in Emilia Romagna would have been much less. So what is the technological profile in the period since the Piore and Sabel data were gathered, and how is work with new technologies, such as CNC, organised?

Costly FMS and even FMCs are beyond the financial resources of almost all artisanal firms. On the other hand, non-integrated computer technologies, such as CNC, CAD, some CNC machining centres and robots, are financially feasible for such firms. Despite the low adoption rates referred to above there is evidence that the more ambitious artisans were committed to using computerised machinery by the mid-1980s. Over 20 per cent of the 120 Bolognese artisans said that they planned to increase or introduce CNC (ECIPAR/FNAM 1984, p. 88). While in the Modena survey over 34 per cent had introduced NC/CNC during 1981–5 and 21.4 per cent were planning such investments in the next two years (Rieser and Franchi 1986, p. 38). Thus while NC/CNC technology could have had only a very minor role in artisanal competitiveness and success in the growth phase, such technology may since have become much more central in some districts and certain enterprises. So what was the 'traditional' form of work organisation and how does this combine with newer technology?

Division of labour and work autonomy

The metalworking artisanal firms had a notably higher general quality of labour than in the larger firms. A comparative study of artisans and large industrial firms showed only 11 per cent more employees with secondary school and professional qualifications in the latter. Despite reports of high labour turnover in the artisanal firms (Capecchi et al. 1978), over 46 per cent planned to stay with their current workshop. Training also seemed to be more extensive with over 50 per cent of artisanal employees claiming that they were still learning their jobs for one year or more, compared to only 30 per cent amongst large-firm employees. (Rieser 1986, pp. 180–1). But the 'traditional' organisation of artisanal work does not seem to vary overall from that in larger industrial firms. In terms of their time management, repetitiveness of tasks, and discretion in varying their procedures there was little difference between the artisanal and industrial employees (Rieser 1986, pp. 188–90).

These findings tend to undermine the flexible specialisation theory that work in the artisanal firms is more open and rewarding than in the large firms (Sabel 1982, pp. 220–1; Piore and Sabel 1984, pp. 273–4). However, these are aggregate verdicts on a variety of quite different employment situations. The fine detail of work tasks may well be more polyvalent than can be assessed by the broad indicators of surveys. Indeed, several of the artisans that I interviewed did believe that they had to be able to offer skilled and interesting work because workers' expectations were no longer confined to adequate wages. Provincial and regional officials of the communist–socialist metalworkers' union, the FIOM, also believed – albeit from a concern at the shift in employment from large industry – that the artisans often offered more satisfying work opportunities than the industrial firms.[7]

However, like some Japanese factories, the most likely pattern is neither generalised polyvalence nor skill polarisation. For example, in a study of 219 Modenese workshops, with a broad spread of skills and experience, individuals gradually built up their skills over time and were expected to transfer to different kinds of machine according to their expertise (Rieser and Franchi 1986, pp. 52–3). A practice which corresponds well with the lengthy training periods in the artisanal enterprises and their large numbers of apprentices; although the latter is a controversial matter in Emilia. It is often alleged that apprentices are employed only as '*ragazzi di bottega*' – errand-boy apprentices doing menial work – because apprentices are not counted in the maximum number of employees needed to maintain legal status as an artisanal enterprise.

There are also variations according to the kind of product and whether the enterprise specialises in particular kinds of sub-contracted components and sub-assemblies (*conto terzi* work), or makes products for the final customer (*conto proprio*). In the second case expertise tends to be built up around particular parts and processes specific to distinct products; what Rieser and Franchi refer to as 'differentiated polyvalence' (Rieser and Franchi 1986, pp. 53–4). At the Bolognese workshop of the Cavalli brothers, making precision medical instruments (as well as, *inter alia*, parts for British sports pistols!), I was told that a wide range of operator skills was needed because otherwise the firm would not have the capability to produce the necessary product changes. Perhaps most significant is the unambiguous datum that artisanal employees are more likely to report directly to the artisanal owner than to an intermediary supervisor or manager, as in the larger firms (Rieser 1986, p. 190). As we shall see in the context of CNC usage this minimisation of hierarchy tends to give much more latitude in work tasks to the employee.

One key difference, in terms of both the task range of the trained worker, and the control and authority exercised by the owner/manager, is the size of the operation. In metalworking machine shops with a maximum of fifteen employees, the artisan can manage the work and involve himself/herself in the more strictly entrepreneurial duties: marketing, raising capital, liaising with customers, purchasing supplies, etc. With more than twenty employees the planning, supervision and co-ordination of production work tends towards a span of control that exceeds the artisan–entrepreneur role. Interviews in the CNA and workshops in Bologna and Modena confirmed that the span of control/size of operation nexus is critical for the division of labour between owner/managers and workers, and within the workforce itself. An exception to this rule is where two or more partners (*soci*) share the entrepreneurial tasks. In one convenient arrangement, for example, one *socio* undertakes the external business tasks, and another the internal design, and production planning and co-ordination work.[8] However, these firms' distinctive flexibility often means tasks are swapped between partners, and between partners and workers, to meet exigencies arising from order to order.

A low number of employees needs more involvement by the entrepreneur in production tasks. To compensate for this distraction from general entrepreneurial tasks, it is more likely that one or more ordinary workers will participate in key production planning tasks: devising a machining operation from a blueprint; recommending tolerances for a specific assembly, or sub-assembly; or just scheduling work flows if the artisan is away on a business trip. Hence there is a virtuous cycle of enforced flexibility: small numbers encourage task sharing, which enhances job satisfaction and capacity to shift between products or processes. Artisans rarely create special grades such as supervisor or foreman to execute managerial authority. Yet at a point in the next size category, varying with product, process or market, the expanded scale of operations leads to special grades of production manager, or supervisor.

In this category also, the employer's expanded range of external tasks necessitates some withdrawal from technical and day-to-day aspects of co-ordination and production planning. A specific category of technical employee is then much more likely. Moreover, the greater size of operations is also likely to formalise the task boundary between these *technici* and supervisors and machinists and other production workers. These size-related distinctions were indicated in the workshop visits, and confirmed in a study of various sizes of Bolognese metalworking firms (Ascari et al. 1984, pp. 17–18). They also resemble the organising of NC/CNC tasks of small owner-managed firms in Britain (Dodgson 1985). The artisanal operation is thus the irreducible, highly successful, essence of the alternative workshop pole of production.

Computerised artisanal workshops

Rieser and Franchi's study of all the larger metalworking enterprises in the Modena CNA covered 219 establishments of which nearly 35 per cent had NC/CNC equipment. The most common form of work organisation in the latter was to allocate one person to do all the tasks: setting-up, operating and programming. Almost always these responsibilities went to more than one worker. The most usual arrangement was for one of the artisan partners and at least one of the workers to alternate on the CNC machine. This pattern may also apply for other computerised equipment. At the workshop of Entimio Verasani in Bologna, for example, I saw how CAD program compilations were shared between the owner and one senior employee. This case conforms to 'differentiated polyvalence'. The CNC tasks are not, and are unlikely to be, split between different individuals and grades. Nor, however, are they likely to be rotated amongst all the workers (Rieser and Franchi 1986, pp. 54–5). Rieser and Franchi speculate that employing specialist technicians might be likely in larger firms if numbers of CNC and programs increased significantly. Yet it was almost never worth the cost of doing this for CNC programming; and market strategies and the 'law of small numbers' would also discourage it.

As the Cavalli brothers told me: market shifts into specialised, higher quality work is a way of avoiding competition. Hence more CNC programming is more varied programming; which must be shared amongst workers and *socii*. Franchi and Rieser confirmed a general trend for more Modenese artisans to work *conto proprio* (Rieser and Franchi 1986, pp. 23–4). But even amongst the *conto terzi* suppliers of simpler components, facing intense competition, this artisanal model of work organisation persists. The Polletti family (father and two sons) make relatively large batches – between 200 to 1,000 – of semi-finished gear parts mainly for tractors and for about ten large-firm customers. Yet they have increasingly pro-rated orders by interlarding these jobs with work for small-firm clients, so as to maintain capacity working. Even here, on the margins as it were of artisanal flexible specialisation, the need to make the most of human resources constrains the Pollettis into polyvalent practices. Their five machinists do programming when required.

Sabel's account notes a skilled–unskilled division in small firms, but does not distinguish between artisanal and non-artisanal firms; nor the detailed distribution of NC and CNC tasks. He identifies a direct role for the artisan in innovative NC programming but not the possible exclusion of manually skilled employees, nor further exclusion of the unskilled (Sabel 1982, pp. 224–5). Greater use of computerised machining could mean two possible developments. One would follow the pattern of the larger firms: retaining the conceptual programming work for artisans as 'managers' or for new categories of technicians. Although, as chapter 3 showed, union pressures checked the transfer of programming tasks from production workers in the larger Italian firms, only a small minority of the artisanal firms are unionised. More programmed production to meet customer requirements for greater routine and precision might minimise workers' present personalised and collaborative contributions. Longer-term results would be redundancies, reductions in skilled workers, and less training: a general deskilling. The second possible outcome would continue the existing bias to polyvalence. Workers could participate in programming, and even take on a task level that larger firms would transfer to technicians or managers.

New evidence shows that an intermediate pattern may be emerging. Slightly more specialisation of technical roles has appeared amongst some Modenese artisans; without apparently increasing the artisans' own influence at the expense of manual workers.[9] Beyond the artisanal threshold changes are more likely. In firms with thirty or more employees overheads are still low. The owner is still likely to undertake several business functions. But formal organisation of functions and specialist occupations are more likely. With CAD and CNC, specialist technicians do product planning and part-programming. The ex-artisanal firm of MG2 just outside Bologna conforms to Piore and Sabel's picture of flexible specialisation. To make packaging machines more closely to customers' specifications it has invested substantially in CAD, CNC and electronic testing equipment. Much production is sub-contracted to networks of artisans in Emilia and beyond. The Bologna plant has fifty employees in production; but their work is highly specialised and hierarchically organised. A production manager supervises CNC and conventional machining, as well as the inspection and assembly. The operators may modify aspects of the program; but this is mostly done by six technican designer-programmers with college diplomas. Operators do not move between different types of machine.

Summary

The artisanal workshop represents a distinct type of flexible specialisation marked by externalising key business functions to collective bodies and internal, polyvalent work organisation. Polyvalence means not so much all workers doing all tasks, as progression through specific jobs, absence of formal supervision, and artisans' proximity to employees' tasks. The regional political culture, and involvement in collective forms of association linked to that culture, may influence Emilian artisans' identification with their workers' aspirations for more satisfying work. However, the 'law of small numbers', imposed by legal limits on the employment size of artisanal firms, also compresses the task range on to a small number of individuals.

Despite the Sabel account, exploitation of NC technology was a very minor factor in artisanal firms' expansion as flexible specialists. In what might be called the 'consolidation phase' of the 1980s artisans in some provinces have extended their flexible capability, either through the *conto proprio* strategy of producing their own products for the final user, or by extending the range of *conto terzi* products and clients which they can switch between. CNC has assisted this evasion of dependence on a limited number of sub-contracted orders; but it is unlikely to change the traditional work organisation pattern. Indeed, as the Cavalli brothers' example shows, CNC may well accentuate 'dis-integrated' production (*decentramento*) throughout the artisanal system, for it enhances productivity without requiring the extra numbers which might exceed the legal limit for artisanal status.

CNC is still a minority of productive capacity both amongst all artisanal firms and within the firms that do possess such technology. Its importance will undoubtedly grow but the source of the artisans' competitive flexibility does not lie in their use of computerised machinery. Checci and Magli – an enterprise of two married couples – assemble and market agricultural devices such as potato hoers, 50 per cent of which they export to European, American and Asian countries. A network of ten artisanal and small firms supply semi-finished units and sub-assemblies; but the couples see no necessity for these to adopt CNC and similar technologies. The strength of their artisanal suppliers lies, they argued, in their ability to solve problems through their organisational flexibility and accumulated experience. Use of CNC and related equipment is according to the traditional artisanal model of work organisation – task sharing between artisans and employees, differentiated polyvalence, and a graduated transfer of workers to more highly skilled work – rather than through new functions, occupations or operating arrangements.[10]

Internal fragmentation of artisanal flexible specialisation, at least in Emilia, thus seems unlikely. Neither technological change, the reactions to economic recession, nor the decline of older artisanal backgrounds seem to reverse their innovative and traditionalist qualities. The artisanal legal form and its social framework discourages most from extending their personal success to pursue independent business growth. Indeed expansion sometimes takes place within the basic forms. Modenese artisans set up 'satellite' enterprises with new partners, but the same internal organisation, so as to run new operations without changing their legal status as artisans (Lazerson n.d., pp. 14, 19–26). A larger threat to small-firm hegemony is, as Bennet Harrison puts it, 'the big firms might strike back' (Harrison 1989). A less Fordist use of FMS-type technologies might reduce small-firm advantages. Medium-sized and larger Italian firms could use FMS to reschedule small batches flexibly, instead of sub-contracting to artisanal and small firms. So the following analysis of FMSs must assess whether such a *recentramento* might also take market share from *conto proprio* artisans; with technological virtuosity producing goods more quickly and flexibly than the artisans can by human versatility and polyvalence (cf. Amin 1988).

We have followed the prospect of computer-integrated flexibility, in the futurist perspective of the cybernated factory, through three countries. The dream of variable production of small batches was only even partially matched by reality in Japan. But might not Italy, where flexible practices seem suffused throughout the social and industrial structure, achieve a more pragmatic version of this paradigm?

FMS and the Italian labour problem

Optimal use of FMS requires a 'production paradigm' consistent with variable production of a shifting range of products, and a correspondingly flexible division of labour. In particular it presupposes a range of skills and a system of work organisation able to provide inventive reprogramming of different products, and responsive correction of software errors and mechanical problems. All-round polyvalence is not essential but those who make part-programs should have first-hand knowledge of machining techniques and problems. Schedule programmers must be in regular contact with the mechanical systems and understand machine capabilities and problems. An overlap of tasks amongst the teams working the FMS is optimal, so that knowledge and skills can be shared within the limits of defined accountability for specific tasks. Polyvalence could take the form of operating staff swapping tasks at the same level of 'horizontal' competence when exigencies required, having a basic understanding of the constituent electronic, software and mechanical systems; as well as being able to take on programming tasks in conjunction with, or in the place of, those who prepare software for new operations and products (see Jones and Scott 1987). The 'Alpha' case, in the USA, closely resembled this model arrangement.

At first sight it seems likely that Italy, and especially firms in the 'Third Italy', could well realise this paradigm. Italian unions, as we saw in chapter 4, made an egalitarian redistribution of skills and the abolition of Tayloristic specialisation one of their central demands in the 1970s. Italian managers have the evidence of smaller-firm, flexible specialisation practices before their very eyes. Indeed in many cases they use such arrangements by sub-contracting some of their own work to artisanal and small-firm suppliers. They also produce for a domestic market in which the interaction of decentralised demand and supply has, for some years, meant diminished batch sizes and a preference for more customised products. Yet once again, the paradigmatically rational is frustrated by the weight of institutionalised practices.

Unions and labour

The egalitarian radicalism of the *professionalità* movement in Italian unions petered out in the face of the employers' offensive for redundancies and job rationalisation in the early 1980s: an experience both stimulated and symbolised by the reassertion of managerial power at FIAT (cf. Cressey and Jones 1991). Elements of this campaign have been refashioned into a more pragmatic work reorganisation policy by Emilian union officials (Marchisio/FIOM 1990). Yet even in Emilia the decisive force of factory unionism has been forced back into defensive bargaining positions in recent years (B. Jones interviews at Sasib 1985; Murray 1984). Union officials see the status of technical workers as a barrier to radical work reorganisation proposals. In larger firms these have expanded in numbers in line with technological change, but in most cases they are not unionised.

In Britain, as we saw earlier, struggles to expand the 'upper' boundaries of production workers' jobs often succeeded where the technicians' union was weak or non-existent; or they involved inter-union contests often with managerially engineered compromises. In the USA industrial relations law and practice normally placed technicians outside of the sphere of union bargaining, and thus precluded unions from aiming at the delegation of their programming and planning tasks. In Japan technicians hardly exist as a distinct occupational status in production functions, and they could anyway be included in the multi-occupational enterprise unions. In Italy the same unionisation principle applies. There are no legal barriers to technicians' unionisation and the industrial rather than occupational basis of membership allows their recruitment. In practice, however, Italian metalworking unions have had very limited success in enrolling technicians (Rieser 1988; FLM-Bologna interviews March 1985). Yet unions still aspire to increase their organisation of technicians. Consequently unions are ambivalent about campaigning for reorganisations of work that will transfer work from technicians and hence increase their reluctance to join the union.

Managerial investment perspectives

The earlier discussion of technology diffusion in Italy indicated some of the problems of FMS adoption. The large-firm/large-batch vs. small-firm/small-batch polarisation of industry leaves a very restricted pool of potential FMS users because FMS, as conventionally marketed, is for medium batches of well-defined part-families. Many FMSs in the bigger firms are used 'rigidly' in US style, for the necessary longer production runs. Mandelli and Comau are aware of the need for further customisation of their FMS products to appeal to the smaller batch users. However, many managers, who are potential purchasers, have little inclination to be risk-taking pioneers when they can get cheap and relatively reliable production from sub-contracting. On the other hand, industry sources claim that there is a growing market for the cheaper and less sophisticated Flexible Manufacturing Cells (FMC). These are essentially CNC machining centres with DNC links to host computers for automatic downloading of programs (Cazzaniga Francesetti et al. 1988, pp. 83–4). In the longer term, however, such an approach will lead to the problem of 'islands of automation'; which will be difficult subsequently to integrate because of incompatible interface and software standards.

According to industry experts, some of the larger Emilian firms that could invest in FMS were postponing decisions on this stage of development. This was either because they saw greater benefits in automating just the design and logistics of their operations. Or, as with the substantial packaging machine sector firms, because the current trend is towards the replacement of mechanical by electronic parts (interview with Giuseppe Chilli, Bologna, July 1985). There are risks that investment in costly metalworking equipment might later prove redundant. A senior manager at one Bologna electro-mechanical machinery maker explained that FMS was not likely in the near future; their considered alternative was 'top-down' progress towards CIM through detailed introduction of CAD–CAM, which would enhance market responsiveness by computerising the design and logistics functions. Despite having a considerable NC/CNC capability since the 1970s this firm felt that there were considerable advantages in continued sub-contracting to artisanal and other small-firm suppliers.

Amongst the Italian firms that have invested in FMS, suppliers found it difficult to identify any predominant aims. Lowered costs, higher quality, faster throughput of parts were all reported in interviews. Investigation of four FMS operations in Emilia, about a quarter of the then population of Italian FMS, also provided a mixture of such motives. The volatility of product markets – a factor not normally reported in US, Japanese and British enquiries – was also stated by these users. In the late 1980s there were three of these Emilian FMSs that were fully operational, with a fourth in the construction stage. In all three there were marked variations in their operating principles; variations which were linked to labour relations and product market factors.

Three Emilian flexible manufacturing systems

Emilian engineering proficiency has had international publicity in the past, because of Grand Prix successes by two high performance car manufacturers. Both of these firms, Velocita and Campione,[11] were privately owned, and had lost competitiveness in recent decades. Both were taken over: Velocita by a combination of state finance and a private entrepreneur; and Campione by merger with a large-volume car manufacturer. Both firms had a reputation for employing highly skilled craftworkers on small-batch, and sometimes one-off, products. The small-batch/high-skill nature of their production made these firms anachronistic non-Fordist car producers. The installation of the FMSs, however, was part of a general shift in their labour and production policies.

At Velocita the new management saw labour control as one of the main problems. In labour relations the trade union was seen as having too much influence and the workforce numbers as excessive. Bargaining was suspended for several years and only restored in 1984. The union saw the introduction of the FMS as part of a general rationalisation of labour and working practices. Employment was reduced from 350 to 200. Union representatives felt that the FMS had taken away the accumulated skills of the production workers. The FMS consisted of two lines which machined different dimensions of engine blocks. All programming was concentrated in the office of the planning engineer and his assistant. Managers thought that the previous manual methods had allowed too much worker control and discretion. The twenty FMS operators, who covered three shifts, were not allowed to modify programs, and each shift was directed by an FMS supervisor. In line with the hypothesis that organisational flexibility will be needed to get FMSs sufficiently adaptable to take back sub-contracted work, management at Velocita felt that this FMS was insufficiently flexible to bring back sub-contracted work.

At Campione wage levels had previously been high and workforce skill levels widely esteemed. The union had maintained a continuing influence during the crisis years of the late 1970s and early 1980s. But they had only a general notification about the introduction of the FMS. Management allowed modification to the FMS programs and the workers were given short courses on the basics of NC machining programs, electronic controls and informatic systems. The system was similar to that at Velocita and the numbers of operators the same. However, there were also eleven technicians employed primarily or mainly on the FMS: two with programming responsibilities, three to handle the electronics, and six supervisors with technician gradings. There were also a departmental supervisor and beneath him a shift leader (*capo di torno*) for each of the three shifts. Reminiscent of the early days at Alpha (see chapter 7), Campione had gone outside the firm to recruit 60 per cent of the technicians and operators for whom Campione was their first job. The average age was only twenty-three.

The FMS was used to make a wider variety of parts – engine blocks, gear-boxes, cylinder heads and clutch plates. However, this wide range of capability had not seriously affected sub-contracting. Less than 10 per cent of previously sub-contracted work had been put on the FMS. Here was a greater level of product flexibility than at Velocita, without, however, an intention to recentralise previously sub-contracted work. Much of the FMS workload was dedicated to parts for a particular series of engine and transmission units that would otherwise have been made on in-house machines and transfer lines. The low trust ethos of Velocita was only partly in evidence at Campione and the elements of task sharing had the imprint of the supplier – Mandelli – who provided training: only, however, for the *tecnici*. More generally the staffing seems top-heavy with technicians, and the team as a whole has a very limited amount of previous machining know-how. No information was available on the reprogramming of the FMS nor on the ratio of down-time. However, on the evidence of FMS operations elsewhere, the organisational structure and experience of the FMS team do not suggest that operating problems and future flexibility will be dealt with easily.

The third Emilian FMS was a large tractor manufacturer trying to adjust to the lowering of batch sizes and fluctuations in demand by replacing a transfer line with an FMS. As a division of a larger vehicles group it was part of the 'big industry' structure in which FMS is aimed at more rigid, quasi-Fordist uses. The plant was mainly for assembly and 'difficult' machining. Most parts came in from a variety of large and small sub-contractors, and some from other divisions of the group. The FMS is not directly relevant to these requirements. Much more FMS investment would be needed to reclaim sub-contracted work, and then there might be new problems with extra stocks of materials and work-in-progress. FMS operators, recruited from elsewhere in the factory according to previous relevant machining experience, were expected to load pallets, monitor the FMS processes and respond to mechanical problems. They could propose program modifications but not actually make them. This was done by a new grade of technician programmer, called the computer supervisor, from a separate computer room like the system supervisors in American FMSs. Supervision of operators was shared between two of the existing production supervisors, in order to save the costs of a specialised supervisor. The management's public position was that neither the union nor the operators had expressed an interest in wider work roles.

Interviews with union representatives gave a different picture. The union had been on the defensive for some years during the recession and crisis years; but it had maintained a proactive stance on work organisation. Consultation was still practised and the union had been informed one year before the installation date of the FMS plan. However, this was only *after* the design of the computer systems had been decided. The union stance was still one stressing *professionalità*. It was particularly opposed to the arbitrary selection and training of FMS personnel by management. The union had proposed more integrated work roles with programming tasks for operators. However, in addition to management antipathy, the union representatives were also concerned about the response of the *tecnici*. The latter's union involvement was either weak or non-existent. Yet union philosophy was universalistic rather than sectional. They wanted to recruit and mobilise more of the technicians and were concerned that a campaign to transfer programming work to operators would alienate the technicians.

Summary

On the basis of the evidence from these three FMSs, and the reports from FMS suppliers and other managers and experts within the machinery industries, the prospects of a computer-integrated displacement of artisanal superiority seems unlikely. It is likely that some FMS plants, such as Campione, would out-perform some individual artisans for output and product range per person. It is highly unlikely that their capital outlay and service infrastructure costs could match the equivalent costs and product range of a co-operating group of artisanal workshops. The division of Italian industry into Fordist and decentralised spheres causes structural and operational barriers to adoption. There are also internal enterprise obstacles to a *recentramento*.

FMSs may be used rigidly, rather than for a wide variety of products. The sheer scale of much of the sub-contracted production would require extensive and highly expensive computer integration to bring back in-house, or to compete with artisans and other small producers. FMS managers' paradigms, like their American counterparts, do not encompass flexible and non-hierarchical work organisation and management systems. Yet without these arrangements the FMSs are going to be difficult to reprogram and adjust for continuous and variable usage. Ironically the one force that might have pushed Italian managers to accept such polyvalence and delegated controls is the work equalisation philosophies of the metalworkers' union. But recession and the reassertion of managerial prerogatives have blunted and displaced this force. Managers are left to run FMSs with considerable technological capacity but moderate operational flexibility, compared to the networks of flexible specialists outside the big factories.

Conclusion

Despite the technological advances in metalworking machines, and the undoubted technical spin-offs from Fordism and Taylorism, an Emilian artisan would soon settle down to working in Boulton and Watt's eighteenth-century manufactory. There is a nucleus of craft technique, which spans this technical trajectory, and which accounts for perhaps half of the competitiveness of these modern craft workshops. The other half of this contemporary success story derives from the collective organisation of individual workshops, an organisation that is decisively rooted in political cultures and social institutions. In this chapter I have tried to confront the technological trajectory of computer-integrated production, evangelised by the discourse of futurism, with its opposite: disintegrated skill-dominated technology in socially regulated units of production. Previous chapters have argued that the CIM-oriented factories have made only partial breaks with Fordist production paradigms. In some cases they amount to little more than a more realistic Fordism for the batch manufacturer: a kind of neo-Fordism. Where a transition to a paradigm resembling flexible specialisation has occurred it has been for contingent reasons or, as in Japan, because of macro-social and external industrial structures.

If flexible specialisation has an essential core then this will not be the *presence* of craft skills – for these may flourish even in the niches of the Fordist factory. Rather it must consist of craft forms of work *organisation*: a capability to vary production between qualitatively different products, and the small production runs that make these skills and variations feasible. In this chapter I have taken the flexible specialisation hypothesis back to its empirical and intellectual source in Emilia Romagna. The core elements are possible in various enterprise forms, but the conclusion here is that only in the artisanal workshop does flexible specialisation have any *necessity*. Furthermore, this is a necessity that is relative to the prevailing socio-political context and to the national industrial structure, rather than absolute. Nevertheless, polyvalence amongst production workers and a sharing of responsibilities with the owner-managers is essential for artisans to maintain individual competitiveness. Variable product capability is likewise necessitated by the need to offer something different from Fordist-style large-batch producers, and partly by the need to make use of pre-existing capacities and aspirations of skilled craftworkers.

Similar organisational profiles exist elsewhere, but only in Italy, and particularly in Emilia Romagna, does a social structure exist to define and reinforce the artisanal organisation of production. Legal conditions, themselves arising from past political contests, sustain and restrict the artisanal activity. In particular the restrictions on numbers employed and the exemptions for the employment of apprentices enforce labour and skill-intensive production, through what I have called 'the law of small numbers'. The representative associations add to the social identity of the *artigiani* and make possible collective forms of interest representation. The speciality of Emilia Romagna consists of the intensified collectivism which the regional political culture imparts to this common interest, and to some egalitarianism in work roles. Unlike other countries, and some other Italian regions, this institutionalised collectivism gives the artisans both a degree of purpose, and a continuity that transcends the particular problems of individual small firms. Many medium-sized and larger firms cannot match the resources which this collective agency provides.

This collective 'corporation without a roof', to use Amin's terminology, thus differs from other, normally larger, types of small firm. These might also practise aspects of flexible specialisation, but they do so without either an essential craft organisation of work, or the permanence of the external structures, which help to protect individual artisans from the forces of economic disruption. The flexible specialisation hypothesis, and its qualified confirmation in this book, could be interpreted as a statement of industrial Darwinism.[12] In other words: flexible specialisation must win out over its more primitive competitor systems. This is untenable. The application of Fordist techniques, and CIM technologies, to metalworking batch production factories can bring some cost and variety advantages in particular environments. But, without support from a changed social organisation, most of these computerised plants will not break from their Fordist, or neo-Fordist, paradigms into flexible specialisation.

Moreover industrial systems, like natural ecologies, are symbiotic. Both the Japanese and Italian examples show how different 'factory' systems can both coexist and support each other. The present analysis supports one major assumption of the Piore and Sabel approach: to show how the inter-penetration of the constituent technological and social trajectories of the factory make for multiple paths of manufacturing evolution. But in this sense Emilian-style flexible specialisation is no more privileged than the factory of the futurist paradigm, currently being forced through other social and political channels. Emilia is the exception which disproves the certainty of their rule. Its potential to become the rule is another matter.

11 Conclusion: the struggle continues

This study of the pathology of the metalworking industry began with three questions about recent changes to production processes. Will computer technologies finally establish true factory principles through cybernation of the 'workshop' sphere of batch production? Is the emerging organisation a continuation of earlier industrial revolutions, or is it qualitatively different? Are the newer practices sufficiently powerful and universal to transcend the national variations that have developed in the past? These questions required both theoretical and substantive analysis of the technological evolution and social structuring of production organisation. The most fundamental theoretical clarifications concerned the meaning of factory organisation, its historical forms, and the relative influence of technological, organisational and socio-political factors.

Probing the debates over the continuity or discontinuity of technological shifts in manufacturing showed the significance of the analytical categories of Fordism and Taylorism. By revising their conceptual relevance to the problems of batch production in metalworking these categories have been refashioned. The substantive evidence concerned national variations in the pursuit of cybernation, the historical application and modification of Fordist and Taylorist principles, and the interaction of socio-political, economic and technological factors in each society. Together these theoretical refinements and empirical assessments have allowed us to redefine the problems, the past and the likely future of production in the metalworking core of manufacturing.

Conceptual and historical clarifications

Batch production: inside or outside the factory logic?

Asking whether batch production fully departed from the workshop form of organisation challenged the idea of a homogeneous factory institution. We saw that the factory system of nineteenth-century industrialisation could not have revolutionised metalworking engineering. Before that period there were no machine-tools, no engineered components – in the contemporary sense – and no industrial process comparable to metalworking engineering. The early nineteenth-century engineers invented their methods and equipment in the course of improving the new machinery products with which they were concerned. The classic 'factory system' of the textile and related industries was never established in metalworking engineering, even as Marx, Babbage and others were hailing its universal necessity. The small size of the batches and the range and complexity of component production and final assembly inhibited factory methods and processes in metalworking. However, operations such as Boulton and Watt's Soho works – allegedly prototypical of modern factory systems – do exemplify the inner contradictions of batch production.

Beyond a minimal size of operation, the specificity of the various constituent processes – casting, metal removal, machining, fitting and final assembly – required two preconditions to simplify and co-ordinate work into an orderly flow. Much of the detail of operations needed complex pre-planning and central control. The other requirement, partly recognised by Boulton and Watt, was to simplify and mechanise operations by making a smaller range of products in larger volumes. But metalworking engineering means batch production and, thus, incompatibilities with these two solutions. Most basic metalworking, particularly before the mid-twentieth century's Fordist mass consumer products, produces capital goods. With historically small and variable markets the nature of the main constituent processes – turning, milling, drilling and fitting – varies considerably with the product. Tools and skills are correspondingly highly particular. Hence the ubiquitous potential to decentralise the operation of constituent processes to a 'pre-factory' level of workshop organisation: either within the main plant, or to small, outside sub-contractors. Even conscious integrators like Boulton and Watt resorted to such sub-contracting arrangements and specialist workers.

Two modes of factory integration: Ford or Taylor?

Standardised integration and pre-planning, together with centralised control, did come with the transformations of technology and organisation personified by Ford and Taylor. The logic of Fordism and Taylorism in the actual practice of their founders revealed two distinct attempts to transform inherent elements of workshop structure into factory levels of control and integration to change batch production. Fordism completed the standardisation of products and processes started by the 'American system' of interchangeable parts with pre-defined dimensions and tolerances in arms manufacture. Ford combined pre-definition of products and process tasks via standardised machining and sequential assembly with maximum integration of activities into single factories. However, this quantum leap along the horizontal axis of production organisation presupposed raising output to very large batch or mass production levels, and minimal product and component changes.

By contrast Taylorism was essentially, though not solely, focused on the vertical dimension of production control. It aimed at central and hierarchical controls and standards, largely by administrative and bureaucratic procedures, without uniform products and processes. These conceptual clarifications are useful in reassessing the significance of Taylorism and Fordism for reorganisations of batch production and more recent moves to automate and cybernate tasks and processes. Much previous analysis of these developments has, I would suggest, been handicapped by insufficiently clear distinctions between Taylorism and Fordism, and inadequate definitions based on characteristics such as task fragmentation, assembly line technology, mass production and deskilling. Having made this distinction we can see that metalworking made very little progress towards factory systems of production. It has relied more on Tayloristic measures than Fordist reorganisations which involve product standardisation and pre-planning of designs with integrated operations. Indeed the 'crisis of Fordism' for metalworking has been how to apply it successfully, rather than how to transcend it.

Production paradigms and national trajectories in Europe

The subsequent history of metalworking batch production consists largely of attempts to extend Fordist and Taylorist methods and techniques; both into metalworking's various functions and sub-sectors, and also overseas. Chapters 2 and 3 showed how the initial bursts of Taylorist and Fordist innovation mellowed in the USA, but were resisted in Britain, initially by indigenous production paradigms and practices. The decisive conjuncture for the export of Fordism and Taylorism, the crisis phase for the indigenous paradigm, was the period of European and Japanese reconstruction after the Second World War. The US Marshall Plan's technical assistance programmes fed European political aspirations for US-style mass-consumption economies and a wide-ranging interest in Tayloristic and Fordist methods and techniques; which in the US had by then become merged into functions such as industrial engineering. Despite major shifts towards standardising product characteristics and hence production functions, major advances in Fordism in Europe were always restricted by the limited size of markets. Tayloristic administrative controls, especially work-studied PBR (payment-by-results) schemes, were seen – especially through British managers' ingrained production paradigm – as the essential modernisation practice.

This sharper distinction between Taylorism and Fordism helps identify the relevance and nature of the institutions and contests making up the social trajectory. In Britain the key social institution was the collective bargaining system, which gradually grew in influence during the post-war drift towards corporatist decision-making practices. Because this was built on the earlier traditions of devolved shopfloor bargaining and representation, unions were able, in the absence of concerted opposition from managements, to adapt Tayloristic PBR to their own expectations; blunting its already limited potential.

By contrast, Italy's social trajectory took a different direction. Stagnation under Fascism, and then military conflict meant even greater technological backwardness. This weakness made crude applications of Taylorism more attractive to managements than detailed and sophisticated adoption of Fordist techniques. The social trajectory of Cold War politics split the labour movement and marginalised adversarial unionism, facilitating the pursuit of vulgar Taylorism. Weak union influence permitted a generally unscrupulous approach to labour utilisation, scientific management meant crude and exploitative intensification of work. This ruthless and expedient dash for higher productivity temporarily boosted Italian manufacturing, but left a legacy of labour grievances and anti-Taylorist sentiments with potent consequences for later periods.

Computerisation: competition between paradigms

In the main Fordism and Taylorism only adapted technologies to pursue logistic and organisational changes. The source of the next major effort to solve batch production complexities developed within the technological trajectory itself: from computerised machining through numerical control systems. Although it played upon predilections for Fordist-style process efficiencies, in the USA and Britain, the managerial conceptualisation of NC took on a predominantly Taylorist bias. Indeed, its consistency with vertical control principles, and potential for piecemeal adoption, seemed to swing batch production priorities back from Fordist techniques and more systematic Fordist devices, such as the rival Group Technology and integrated cell systems. Receptiveness to NC was thus facilitated by the persistence of a Tayloristic production paradigm amongst plant managers. However, the decisive factor favouring its wider adoption was the decidedly non-Taylorist compatibility of NC with shopfloor skills. The strength of low-trust Taylorism and stable institutions of labour regulation made this accommodation largely tacit in the USA. In Britain and Italy, on the other hand, the growth phase of NC, and especially CNC, coincided with more active workplace unionism in the social trajectory. The realisation of Taylorist organisation through NC/CNC was therefore checked.

As a deskilling device, NC also had greater technical limitations than is recognised by either its academic critics or some of its industrial proponents. However, the resulting scope for continuation of shopfloor skills was extended further, into the sphere of programming tasks, by workplace unionism. British and Italian workers have probably gained more influence, in this respect, than their counterparts in the USA. The two European nations' labour relations institutions were still in a state of dynamic flux during NC's growth phase. However, in the USA both unions and managements were circumscribed by a relatively static system of contractual negotiation around individual responsibilities and rights: restraints which maintained a more rigid division of control. In Britain, and even more in Italy, an upsurge of shopfloor militancy and bargaining power underpinned machinists' influence over NC/CNC programming functions. But the sources and unevenness of this influence were determined by different industrial relations trajectories.

The unprecedented power which workplace unionism achieved in Italy in the 1970s was partly caused by the primitive and unilateral Taylorism of managements in the immediate post-war period. But the capacity of this new unionism for detailed reorganisation of NC/CNC tasks was limited by ideological preoccupations with an egalitarian division of labour, and with political anxieties about alienating the under-unionised technicians' responsibilities for programming. In Britain, however, a roughly similar level of worker control over NC arose from resurgent shopfloor bargaining based on revitalised craft bargaining tactics, openly competing with technicians' prerogatives, but lacking any overall strategic perspective.

By examining the origins and diffusion of NC in chapters 4 and 5 it has been possible to show both the boundaries of technical and social influences, and the general parameters involved in the broader computerisation trajectory. Technologists such as Williamson clearly recognised the essentially particularistic quasi-workshop character of small-batch production. But their solutions, of integrated cells and Group Technology, were essentially schemes for Fordist standardised integration, which ignored the continuing appeal of Tayloristic hierarchical piecemeal controls. NC seemed to win out over these quasi-Fordist systems because of its initial appeal to Taylorist expectations. That slice of technological history puts this book's main focus – on the putative transformation of small-batch production by cybernetics into the 'Factory of the Future' – into its proper perspective: an inherent tension between decentralisation to constituent processes and rationalisation through either standardised integration or bureaucratic hierarchies of control.

The future: forced or fragmenting?

The clarification of the relationship between small-batch production, factory and workshop forms of organisation and the distinction between Taylorism and Fordism has been combined with the concepts of production paradigms and national social and technological trajectories. Applying this framework to the continuing evolution of production systems has cleared the ground for answers to the initial questions. Those concerned the prospects for a fully cybernated batch production through computerisation, the likelihood of a universal model transcending previous paradigms and constraints of national particularity, and the related intellectual problem of whether the emerging pattern(s) deepens and continues, or breaks with previous lines of development. These issues are closely interlinked, but let us consider first the implications of the differences indicated between trans-national diffusion of universal flexible automation technologies and national variations in their use.

When industrial and state agencies pursued the elevation of computer-controlled processes into technologies such as FMS, the genuine potential for unprecedented product versatility was overlaid by preoccupations with bringing Fordist levels of integration and pre-specified operations to batch production. However, the multiple potentialities of the systems brought the latent contradictions in production paradigms – between Taylorist bureaucracy, Fordist integration and pressures for product versatility – into manifest conflict. Japanese firms have been better at avoiding these contradictions; not because of inherently superior technological details, nor because of greater managerial expertise. The principal facilitator has been the more adaptive work roles and employment relationships, rooted in Japan's specific social trajectory.

Japan: post-Fordist exemplar or national hybrid?

There is no universal logic to the operation of FMSs in Japan. Fordist-style use to gain economies of time, labour and capital utilisation is common. Yet other plants aim at versatile operation for a wide product range; with much reprogramming, involving different grades of employee. However,it would be a mistake to follow the interpretation of Piore and Sabel and equate this form of work organisation with the craft techniques of the skilled artisan's workshop. It is also misleading to suppose that either Japanese managers, or Japanese work organisation, is uniquely suited to flexible automation. With virtually no European-style craftworkers in large Japanese manufacturing firms, nor a distinct grade of technicians, the workgroup is made up of workers with varying degrees of occupational and skill status. Skills are imparted on-the-job, mainly in learning-by-doing mode, and only as and when needed. Skills correspond to experience; and experience and authority tend to reside with the more senior workers at, or just below, supervisory level. This role encompasses the craft, technical and supervisory functions that Western, Tayloristic systems assign to separate occupational grades. A capacity to transfer know-how from a specific department and workgroup to another workplace is a hallmark of traditional European craftworkers. In Japan, only the longer-serving, most technically informed senior grades of worker are likely to have the breadth and depth of skills to accomplish this transfer easily.

In applying these inputs of labour Japanese managers are not showing any great strategic prescience to exploit new technology. They are simply utilising the occupational and organisational practices that most larger firms inherited from the socio-political trajectory of the 1950s and 1960s: the conflicts and compromises which institutionalised lifetime employment, seniority-based employment systems, and enterprise unionism. Within the limits of on-the-job and incremental learning, the resultant system of work organisation is elastic. It can run on minimal expertise under simplified production goals and processes, or it can be stretched to increase the mean levels of skill, within the bounds of the formal engineering knowledge and senior production workers' accumulated experience available in the firm. Unlike the USA's contractual individualism, the pattern of skills and work roles hinges on a key principle of labour regulation, a 'meritocratic cohesion'.

Typically, production workers in medium to large Japanese establishments have the promise, but not the guarantee, of incremental promotion on the basis of proven experience and seniority to higher wages and positions within a production department hierarchy. In Western, Tayloristic, companies the middle ranges of the occupational hierarchy are outside the production areas. In Japan such hierarchy is more likely within the production and related functions. The perceived opportunities of promotion within the hierarchy of production grades provides an incentive to individuals to co-operate and learn for themselves in the problem-solving activities on the shopfloor. The system also obviates the need for Tayloristic styles of external controls and hierarchies of specialised roles. However, while it is non-Taylorist, the adaptive workgroup in flexible manufacturing systems arises from social dynamics rather than managerial strategy or a novel production paradigm.

On the evidence of the machinery sector, at least, while there is a post-Taylorist – or more strictly a non-Taylorist – work organisation, the production paradigm would be hard to classify as post-Fordist. Product standardisation and pre-planning are consistent with a Fordist perspective, although the element of, at least, potential flexibility stemming from labour organisation suggests a modified or neo-Fordist label. The general reserves of artisanal-type skills that can be utilised by sub-contracting to the small-firm sector provide a useful alternative source of manufacturing expertise.

Constraints on computerised flexibility: national continuities

By contrast, in the USA, the industrial relations trajectory has, so far, reinforced Taylorist production paradigms. Managers struggle just to use their high-tech flexible automation investments for more limited neo-Fordist goals, or aspects of traditional Fordism. Flexible automation in the USA has been nourished by the ideology of futurism: the introjection into current technological planning of an idealised version of future automaticity. Some technologists hoped that systems like FMS would lead to variable production of distinct small batches. However, the absorption of five decades of Fordist principles into US manufacturing made quantitative economies in labour, lead-times and machine times the most appealing to US managements. Into this restrictive paradigm was interwoven Tayloristic norms of work organisation and labour regulation. The industrial relations system of contractual individualism, segregates job descriptions and helps to maintain manual skills during the phase of NC diffusion. However, with FMS types of installation the tight job boundaries and the iron curtain between manual unionised and technical-managerial jobs has reinforced the anti-labour and anti-versatility biases inhering in US managers' approach.

Some IR experts in the USA have claimed that a relaxation of collectively bargained job segregation is under way (Kochan et al. 1986). However, contrasting uses of FMS in the case studies labelled Machinery and Alpha indicate that such modifications are likely to have mixed production outcomes. The former plant was a relocation partly to evade unionisation. Yet managers retained individualised and segregated job descriptions as bedrock assumptions of their Tayloristic preconceptions. By contrast, Alpha's partial avoidance of seniority provisions and tight job descriptions coincided with business pressures for versatile and pragmatic use of the FMS. This mix of circumstances allowed fluidity in skills and responsibilities amongst the workgroup, and a paradigmatic case of flexible specialisation. However, the key condition for this flexible specialisation was that the FMS workgroup effectively took control of the running of the system; because of their relative independence from detailed union regulation, and from managerial authority. As an exception to the general rule in the USA, the Alpha case shows more clearly than the Japanese examples that optimising flexible automation's potential for product versatility, rather than cost savings, depends on fluid and relatively autonomous inputs of skills.

The less than innovative continuity with Taylorist and Fordist practices in most US cases might be partly due to continuing stability in the industrial relations trajectory. In Britain, where conditions have not been so stable, FMS and computer integration might be expected to allow a qualitative shift away from previous production practices. Unlike the USA, work organisation and operating strategies have exhibited a wide range of combinations. Shopfloor-bargained craft controls, which had influenced the retention of skilled tasks with NC/CNC, performed a similar – and indeed more effective – function to the job classification/seniority rules in North America. FMS diffusion in the 1980s, however, took place within a different socio-political trajectory. Corporatism had been displaced by economic and industrial policies that had undermined union power at national and local levels and gave managements a degree of power unknown for decades. However, this situation could best be construed not as the management of uncertainty, but the uncertainty of management.

In common with the wider pattern of organisational and industrial relations change, managers in FMS plants rarely had a clear-cut scheme of work organisation and labour regulation to replace the old combination of craft job controls and quasi-corporatist industrial relations procedures. At the two extremes there were deliberate schemes for Tayloristic deskilling and collective polyvalence – the clearest case of the latter being consciously imitative of an assumed Japanese model. In all other cases, however, arrangements were piecemeal, *ad hoc*, and improvised. Jobs retained large areas of craft autonomy, with some cases of renewed Tayloristic controls coexisting with the exercise of craft skills. Perhaps feeling secure that the link between task control and union militancy had been broken, some firms had tacitly ceded programming responsibilities to the craft-trained operators in the FMS, without any formal reclassification of their roles. The British penchant for compromise, ambiguity and informality was clearly manifest in the majority of cases. The eclectic purposes and usage of the FMS were mirrored in the pragmatic forms of work organisation.

Paradoxically, Noble's characterisation of a conspiratorial diffusion of flexible automation in the USA, is more plausible for Britain. Government industrial policies and investment incentives were responsible for most of the early take-up of FMS in Britain. This 'revolution from above' had a productivist bias inasmuch as it strengthened the claims of production managers and engineers against accountants' financial controls (see also Lee 1992). On the other hand, it had the ironical effect of leaving these managers with powerful and potentially versatile systems which did not necessarily correspond with clear-cut product and production strategies. Firms which, like Alpha in the USA, realised some of the potential versatility of an FMS were notable exceptions. Most 'played safe' and saw the gains in terms of Fordist criteria, or possibly restricted neo-Fordist gains such as shortened lead-times and tighter quality control.

The evidence from Italy corroborates the US, Japanese and British indicators on the ambiguities of flexible automation. FMS, its most advanced technical application, is also not being used in Italy to revolutionise Taylorist and Fordist tenets. Italian FMSs are not aimed at full transformation to a post-Fordist logic of variable production of distinctively different items. Once more, Fordist and Tayloristic paradigms persist with the attendant orientation to narrowing and controlling the role of workers. These circumstances limit both the business strategic and flexible labour conditions necessary for something like flexible specialisation to take hold in Italy's 'primary' large-firm industrial sector.

Even more than Japan, Italy's industrial dualism consigns flexibility, as versatile product capability, to the small-firm sector. Flexible specialisation flourishes there, most notably in the artisanal firms; and not, *pace* Piore and Sabel, because of technology, nor only because of craft skills. The key source is a history of politically inspired social alliances, fused with culturally and politically induced enterprise forms, and complexes of representative institutions and services. In brief, whereas the large-scale business pursues flexible automation like its other activities, by abstracting their operation from the wider society, artisanal operations depend on *societal integration* with relevant institutions (cf. Brusco 1982; and for a critical update Franchi and Rieser 1991).

The questions re-examined

Continuity vs. discontinuity

As with other social and economic developments, industrial change interpretations divide into continuity and transformation perspectives. In the latter, history is discontinuous and new developments displace most or all of the preceding arrangements. For continuists, on the other hand, close inspection always reveals some persistent underlying logic in practices, institutions and relationships. As Smith (1991) has argued, the study of industrial and technological change also features such contrasting views (Piore and Sabel 1984; Shaiken 1984; Williams et al. 1992; Penn and Wigzell n.d.; Pollert 1991). Our examination of new factory systems has tried to reconcile continuity vs. transformation preoccupations by testing the following propositions. Firstly, there is an underlying continuity in the organisation of batch production metalworking which can be traced back to the Industrial Revolution. However, what has been continuous is the tension between the independent technical viability of the constituent processes and the incentives for their overall integration and standardisation. Batch production generally has a double potential: either for decentralisation into workshop levels of organisation, or for standardising products and processes with integration or centralised controls. Therefore, there is a permanent *potentiality* for both continuity and attempts at transformation.

In the early decades of the twentieth century, transformation threatened with the hierarchical centralisation of Taylorism, and the product/process standardisation and integration of Fordism. But these were not complete solutions to the enduring contradictions in batch production. Standardised integration, Fordism, depended on mass markets for single, or very similar products. Centralised hierarchical controls depended on the imposed routinisation of tasks and became identified with the use of work study and financial incentives to stimulate output. Fordism and Taylorism became complementary paradigms in the mid-century production practices exported from the USA. But they were never fully realised in batch production. Hence the propagation of computerised automation in the final decades of this century has been conceived, marketed and purchased on the basis of a contradictory mix of Taylorist (NC), Fordist (FMS) and workshop (CNC) attributes.

For many larger firms the older paradigms have persisted and they use the technologies accordingly. Others, perhaps swayed by the ideology of futurism, see their automated operations in neo-Fordist terms. On top of the conventional list of cost-savings, neo-Fordist firms place priorities such as reduced lead-times, shorter overall production cycles and minimal work-in-progress. Contrary to industrialists' hype, these developments represent continuities with Fordism; rather than the transformation read into them in concepts such as flexible specialisation. Indeed, as chapter 2 noted, in its heroic phase Ford's original automation strategy obsessively pursued some of these techniques and objectives. As larger firms experiment with more realistic forms of organising computer-integrated production it is even possible to find attempts to combine cybernated Fordism and workshop principles. According to Miller and O'Leary, for example, the Caterpillar company discussed in chapter 6 has planned semi-autonomous, simulated workshop cells, controlled and integrated by financial targets and computerised production information systems (Miller and O'Leary 1994).

Universal model or national particularities?

If these continuities are evident are they contributing to a universal model of factory organisation and technology? Here we must reconcile different levels of abstraction. Managements in different countries do share a core of common perceptions in the production paradigms which guide investment and initial implementation. At a more detailed level, important national and local variations in procedures, work practices and the organisation of work roles have been clearly evident, despite the ascendancy of versions of Fordism and Taylorism in most of the industrial countries; at least until Japanese firms adapted North American methods in their own distinctive manner (Maurice et al. 1986; Gallie 1978; Whittaker 1990; Sorge et al. 1983). Flexible automation has some universal characteristics: minimal complements of manual labour and heavy concentrations of internationally marketed software and hardware. British firms buy from US suppliers, North Americans from Italian and Japanese vendors. Yet the evidence on FMS was that national differences are neither eliminated, nor reduced to insignificance.

Apart from common core assumptions about pre-planned, software-integrated, and virtually automatic production processes, there is now no ubiquity of either flexible specialisation, or Fordism, or neo-Fordism, or

Taylorism. The main national differences are between Europe and the USA on the one hand, and Japan on the other. This is essentially because of the successful Japanese revision of Fordism, to what we have termed neo-Fordism, and because Taylorism is absent from Japanese forms of work organisation; at least in batch production. There are also differences amongst the Western countries. Mixtures of Fordism and neo-Fordism predominate, but US users show less variety in their operating goals and remain locked into Taylorist organisational controls. Italy shows slightly more variety of approaches, although the small number of installations on which there is evidence makes conclusive assessments difficult.

In Britain there are wide variations in aims and operational methods. Such variations suggest a production paradigm that is less homogeneous than the Fordist–Taylorist perspective in North America. Britain also highlights another question mark against the idea of a new, uniform system of automated batch production. That is the range of different methods and organisational arrangements that can be found even within the same country. National institutions for regulating the supply, training and representation of labour, as well as bargaining arrangements, set general conditions and constraints. However, the specific product sector, the market location, business strategy, enterprise culture and recent labour relations history, all combine together to cause potentially particularistic ways of using and organising the operation of new systems. In a situation where the national institutional infrastructure is weaker, or in flux – as, increasingly, has been the case in Britain since 1980 – and where managements extemporise to meet continuing volatility in the market environment, heterogeneity and 'plant particularism' are much more likely (cf. Rose and Jones 1985).

Cybernated factories through CIM?

The ultimate constraint on the achievement of cybernation is the continuing importance of human skills. Although these are interwoven into the workshop form of organisation, attempts at factory-like systems of small-batch production never completely eliminated them. In the latter most existing studies (cf. the review in Jones 1988) suggest that some manual manipulative skills are redundant; but this loss is more than compensated by a continuing, though less consistent, need for combinations of conceptual-analytic and craft-mechanical know-how. With FMSs used to achieve Fordist or neo-Fordist aims, the roles of manual production workers – misleadingly often still called 'operators' – can be officially restricted to remedial maintenance. It is true that, even in this instance, the complexity of the equipment – which combines mechanical, electronic, hydraulic and computing systems – requires a range of skills well beyond that of the traditional mechanically trained craft-style worker. However, low product variety and minimal reprogramming, restrict on-going and unplanned interventions required from operating staff.

Yet the distribution, and hence individual possession, of skills is not a simple expression of the technology and the operating strategy. Human skills are the factor which mediates between the technological and social worlds. They can only be realised within the constraints of a given technology. Yet their sources, potency and distribution arise from social conditions and processes. Skill distribution is affected by the production and organisational paradigms, and industrial relations practices. Thus at one extreme there is the combination of managerial Taylorism and collectively bargained job classifications in many US plants. Typically this would mean segregation into low-skilled jobs such as the loaders, limited mechanical adjustments by 'operators', electro-mechanical maintenance and correction by 'repairmen'; with technicians and engineers tackling problems at the interfaces between electronics and computing systems.

Yet at other points within the Fordist/neo-Fordist production paradigm there can be other work organisations, where national institutions of labour regulation, or plant particularism, allow more broadly skilled work roles. A few British and Italian cases did combine mechanical and other remedial skills in what might be called all-round 'neo-craft' roles. However, the major constraint on skill expansion remains the prohibition of programming from these jobs. Again, where Taylorist arrangements were absent or relaxed, Fordism/neo-Fordism did not prevent some minor programming checks and interventions by 'operators': a possibility confirmed by the commonplace involvement of some Japanese FMS crews in checking and modifying both scheduling and part-programs. Higher-level and autonomous skills and tasks in programming on the shopfloor are more likely where a different production paradigm and product strategy – with less hierarchical labour regulation – transcends the Fordist/neo-Fordist paradigm and relaxes bureaucratic and adversarial industrial relations. Alpha exemplifies this possibility. Its production policy of flexible specialisation, plus propitious plant politics, almost dictated programming and electro-mechanical polyvalence amongst key operators. Unlike Japan and the USA, cybernation in European – or at least British – plants continues to rest on some craft skills typical of workshop practices.

The fervour for computer integration has increasingly lost appeal and credibility since the 1980s (Bessant et al. 1992, p. 6). Technologists are openly questioning the flexibility value of CIM and FMS, compared to simpler forms of computer support (Wainwright et al. 1992 p. 122), or Tayloristic measures (Newman and Bell 1992, p. 285; McIntosh et al. 1994). Older Fordist prescriptions for batch production like integrated cells and Group Technology are being readopted (Burns and Backhouse 1994). Indeed the alternative craze for 'Lean Production' seems to reinvent Ford's classic organisational practice (Williams et al. 1992); as well as being hailed, like CIM ten years before, as the final solution to the mass production–batch production dichotomy (Womack et al. 1990 p. 278).[1] A business press survey of 'Computers in Manufacturing' admitted that there was probably not one true CIM installation in the whole world, and described as near impossible the integration of multiple systems within the same factory. Complexities of software integration and high prices were, instead, said to be influencing firms to opt for piecemeal investment in discrete manufacturing cells. A year later the same publication managed to avoid any reference to FMS; and just one passing reference to CIM (*Financial Times*, 24 September 1992). One international survey of US, European and Japanese firms, reported 'no one was getting much out of' FMS and CIM (*The Engineer*, 21 November 1991).

However, the immaturity of cybernation is not diminishing the strength of the business policies and production paradigms that are dictating renewed attempts to introduce Fordism, or neo-Fordist methods, into batch production. The cybernation crusade is still significant for revealing the managerial preferences that guide work organisation choices. 'Flexible automation', for many large firms, means standardised integration, 'top-down' controls and either the elimination of traditional skills, or the avoidance of high skill combinations of programming and mechanical skills. It is either the continuation of Fordism by other means, or greater responsiveness of production to design and market time pressures – neo-Fordism.

Policies for posterity

Let's allow ourselves to end on a Utopian note. Imagine a hypothetical industrial policy dictator, concerned to build up, or rebuild, a manufacturing capability that was less susceptible to the wilder vagaries of market forces, and more disposed to achieve manufacturing competitiveness by combining stable and skilful employment. Allow such an individual the capability to play God a little with the future of batch manufacturing in his/her country. The advice from the preceding evidence would point to possible options. One might be an imitation of Japanese Fordism: reconstructing employee relations around meritocratic commitment and heterogeneous mixtures of on-the-job skills, exploiting the technology for cost-minimising, intensively utilised operations. The other possibility for skill enhancement and market flexibility, would be more likely to lie at the process-specific, decentralised workshop pole of batch production: with 'secondary sector' networks of artisanal-type enterprise; as Mike Best (Best 1992, ch. 9) has indicated in relation to enterprise organisation.

Recreation of the Japanese option is problematic. Although individual Japanese firms seem to have approximated it in some Western plants,[2] a general replication would presuppose either a compact with strongly embedded trade unions, or their virtual negation at plant level. The implications of leaving generation of skills and employment opportunities to companies might also conflict with other aspects of Western nations' labour markets. Other complicating requirements would include a stock of accumulated Japanese managerial experience and, perhaps, also secondary tiers of dedicated, and dependent, small-firm suppliers. By contrast, decentralised workshop networks would not need to cling to, or chase further Fordist levels of output and standardisation.

On the whole it is they rather than large-firm adherents to technological exploitation, that are more likely to possess the capacity to align market demands with skilful use of appropriately advanced technologies; avoiding an obsessive pursuit of CIM. Even so, we would have to accept that even an omnipotent policy maker cannot easily change subjective perspectives. Even sceptics of the CIM reality continue with proposals for its future vindication (cf. Bessant et al. 1992). The history of batch production suggests that paradigms tend not to change overnight in a monolithic fashion. Contrary to the claims of both continuist and discontinuist theories, computerisation has not transcended the ancient dichotomy between factory and workshop. It has merely renewed the strengths which practitioners of both organisational forms can emphasise. Even our industry Leviathan would need to accept that the power and prejudices of explicit and unconscious Fordists will continue. Even were flexible specialists, chiefly in the workshop genre, to achieve widespread success, latter-day Fordists will earn just enough success of their own to merit continuing their schemes; enough to justify their determination, to force the future.

Appendix
Sources and methods

Case studies of NC/CNC users

The primary data used in this book come from four main sources: case study interviewing in British and US firms using NC and CNC technology; case studies of US firms using FMS; interviews with managers, workers, trade union representatives and relevant experts in Japan; case studies, interviewing, documentary analysis and observation in, and of, Italian artisanal and FMS-user firms; involvement in the supervision and data collection for Ph.D projects on British manufacturing automation.

Some of the NC/CNC evidence dates back to the years 1978–81. During that period grants from the then Social Science Research Council (Grant no.: HR 5101/1) and the Nuffield Foundation (Soc/181 (862)) financed case studies of firms using NC and CNC in Britain and the USA. Their support is gratefully acknowledged. In these case studies, of mainly large-firm users, interviews were conducted with managers, workers and trade unionists in Britain and managers and some trade union representatives mainly in North American aerospace plants; totalling around fifty interviews in Britain, and thirty in the USA, from six and five firms respectively (see Jones 1982, 1985). As far as resources and interviewee co-operation would allow, the method of investigation in these studies was based upon cross-checking semi-structured interviews between management, union and worker respondents.

Overseas case studies of FMS: Italy, Japan and USA

This form of collecting and corroborating data was repeated, wherever possible, in the phase of FMS studies financed by the Economic and Social Research Council (Grant no.: F00232144); for whose support I am, again, grateful. This investigation began in 1984 with one of two fieldwork visits to the USA. The US fieldwork focused on nine FMS user firms chosen from a broader pool of about twenty sites reported in the technical and business press or recommended through academic and business contacts. These nine plants covered the following spread of sectors: agricultural vehicles (2 firms); aerospace (2 firms); machine-tools (2 firms); earth-moving equipment, railroad vehicles, and heavy haulage vehicles. They were located in California, Illinois, Iowa, Kentucky, New York State, Tennessee and Texas. The union, management and worker interviews in the plant visits were supplemented by further interviews and documentary analysis with trade union head offices, vendor firms, some corporate headquarters and specialist academics. A total of seventy-five interviews were conducted in the US fieldwork.

The Italian fieldwork was spread over a number of visits to that country most of which were based in Bologna, during spells as a visiting research fellow in the University's Department of Sociology. The earlier interviews were conducted in six artisanal firms, where artisanal partners and sometimes employees were interviewed. These interview responses were augmented by data from a questionnaire survey of Modenese artisans conducted by Vittorio Rieser and Maura Franchi, which included questions proposed by myself. In the second phase of Italian fieldwork interviews were conducted with managers, some workers, and relevant union representatives at three large-firm FMS sites, and one large-firm user of CNC equipment. These representative investigations were supplemented by many interviews with union officials, academic and industrial engineers, plus managers at vendor firms.

Shortage of time and finances did not permit an extensive investigation of Japanese FMS users. For these reasons most of the firms used were both vendors of computerised automation systems and users of FMS. These firms were selected from reports in the technical press which also furnished useful background, and cross-checking sources on the uses, scope and resources involved. For comparative yardsticks I interviewed at two other firms. These were: a small independent CNC-based machine shop with 150 machines, which was also planning an FMS, and a small industrial machinery supplier using mainly conventional, plus two CNC, machine-tools. Supplementary interviewing of full-time union officials, specialist academics and civil servants brought the total numbers of Japanese interviews to thirty. All interviews were either conducted directly in English or through a Japanese academic knowledgeable in the subject area. Interviews in the FMS firms usually involved one senior and one production manager, plus a shopfloor worker where these were willing to co-operate. A deficiency in these Japanese case studies is the absence of independent interviews with representatives of the companies' trade unions. I was advised such an approach would be considered discourteous by the co-operating managers.

British FMS case studies

Most evidence on British FMSs was collected indirectly through research I supervised by three Bath University doctoral students: Peter Scott, Bill Lee and Michael Kelleher.* Their respective topics were craft skills, accounting controls in FMS investments, and skill shortages in high technology firms. Their data came mainly from interviews with key personnel and documentary analysis in separate panels of case study firms. For each of these students I was involved in one or more interviews at the FMS plants they studied. Kelleher's study included only one FMS site but since this was the only British plant of a Japanese machine-tool manufacture it provided a useful contrast with other British and US-owned installations, and a source of comparison with the FMSs I had visited in Japan. Scott's sixteen FMSs are the main source of the data discussed in chapter 9. Lee's sample contained these firms plus six additional sites. I am grateful to Peter, Bill and Mike for subsequent discussions on the salience of their data.

Documentary sources

For some of the chapters additional documentary analysis was undertaken. For chapter 2 the correspondence and employment records of Boulton and Watt, held in the Birmingham Public Library, were consulted. The

* Peter Scott (1987) 'Craft Skills in Flexible Manufacturing Systems', University of Bath Ph.D dissertation; Bill Lee (1992), 'The Cost Conflicts of Flexible Manufacturing Systems', University of Bath Ph.D dissertation; Michael Kelleher (1993), 'Skill Shortages and Work Organisation in British Manufacturing', University of Bath Ph.D dissertation.

information on the technical assistance programme of the Marshall Plan was taken from various contemporary reports such as those of the Anglo-American Productivity Council. I am grateful to Jonathan Zeitlin for identifying these sources for me. Other key documentary sources included surveys by: the Amalgamated Engineering Union (as it was then); *Metalworking Production* magazine (who kindly supplied me with a personal copy of their British machine-tool sources to use). Vittorio Rieser, Maura Franchi and the Assessorato Industria of the Emilia Romagna regional government; and Giuseppe della Rocca and the Olivetti Foundation provided key documents in Italy.

Notes

1 Introduction: explaining factory evolution

1 It is galling to add another neologism at the outset of a book that is littered with those already in use. Yet this is the closest term that is more than the conventional notion of mechanisation of discrete human actions; but is also more concretely related to the engineer's paradigm than the vaguer terms of systemising, or programming. 'Cybernation' is simply a nominative form of a verb, already in general use.
2 Kaplinsky uses the term 'inter-sphere automation' (Kaplinsky 1984).
3 For refutations relevant to the present study cf. *inter alia* Jones 1982; Jones 1989a, 1989b; Wood 1989, pp. 9–11.
4 McLoughlin and Clark also provide an extended discussion of ways of analysing technological effects on work organisation.
5 I am grateful to Jay Gershuny for suggesting this term.
6 A similar criticism has been levelled by Mike Best (1992, pp. 25–6) who argues that analytical priority should be given to the diversity of forms of enterprise organisation. The emphasis in the present book, on the analysis of factory and related forms of production organisation, can therefore be regarded as complementary to Best's approach.
7 *Capital*, vol. I, p. 381.
8 Kern and Schumann 1984; Sabel 1982, ch. 5.
9 For a clear and concise explanation of the practical evolution of cybernetic

principles within manufacturing the reader should consult the excellent account in Hirschhorn 1984.

10 What makes futurist analysis dubious is its conflation of logical compatibility with developmental necessity. Just as Toffler is correct in identifying similarities between more pluralistic life styles and social identities, and diversified manufacture of customised products from decentralised high-technology units of production (Toffler 1981, pp. 189–219), so the factory futurists are correct to point out compatibilities between computerised design, reprogrammable production machines and the automation of human decisions. However, both forecasts are deficient in ignoring the necessity of appropriate social developments to secure the practical integration of the logically related technological and economic elements. Clearly, the kind of new social and economic structure prophesied by Toffler would need new frameworks of institutionalised attitudes and procedures. It would also require felicitous outcomes to the processes of conflict over disrupted social interests, which always accompany change from onc industrial and economic system to another. Finally, the realisation of a new industrial system would require the elimination of the many national differences which, to take the most apposite example, led to the substance of industrial capitalism, varying so much between cases like the USA, Sweden, Japan and Germany. The 'factory of the future' perspective has a much more circumscribed focus than Toffler's societal transformation. Nevertheless its technological viability hinges upon appropriate organisational arrangements which, in turn, depend upon particular aspects of the same kind of institutional frameworks, social processes of conflict resolution, and national differences. In the following chapters these factors will be examined by comparing, in quasi-historical perspective, the progress of technological, organisational and broader socio-political institutions in four different societies. For convenience these will be grouped into trajectories of technological and socio-institutional development, and their implications for past, contemporary and future factory organisation, examined by looking also at the interactions between the different trajectories of change.

11 These were entitled: 'The Definition and Control of Occupational Tasks in the Process of Technological Change' (Social Science Research Council 1977–9); 'Different National Constraints on Use of New Production Technology' (Nuffield Foundation 1982); 'Work Opportunities in Factory Automation' (Economic & Social Research Council 1984–5). Further joint enquiries were made in various post-graduate studentships awarded by the Science and Engineering Research Council, during the years 1982 to 1990. The vital financial support of these bodies is gratefully acknowledged. For more details on data sources and methods of investigation see the appendix.

2 *Past production paradigms: the workshop, Taylorism and Fordism*

1 Source: records of employment and apprenticeships recorded in the Boulton and Watt archive, Birmingham Public Library.

2 In one case the new tooling made such a physical demand on the rest of the machine that the gains from faster cutting were lost through excessive breakdown costs.

3 Taylor's 1910 lectures in Detroit appear to have converted Packard managers to wholesale adoption of time and motion techniques. The evidence that Ford was as enthusiastic is dubious. As Lacey Ford comments on this episode: 'the Ford system, however, [was] to go beyond Taylor, whose "science" was focussed on the time that one man took to do a job. Why use a man at all, was the Ford question, if a machine could do the job instead?' (Lacey Ford 1986, p. 38).

4 By 1954 the new cylinder block line for Ford's Mercury model had 78 per cent of jobs classified as 'set up' or 'patrol' (Bright 1950, p. 100).

5 Zeitlin 1987 argues that British resistance to the 'American System' perfected by Ford lay less in objections to pre-specification of invariant dimensions than in the persistence of varied product ranges and multiple small orders.

Overview to Part II

1 A distinction borrowed from Gould's argument that organisms rarely embody singularly superior anatomies which automatically guarantee their survival and success. See Gould 1991, pp. 234–9, 282–91.

4 Technological evolution and the pathology of batch production

1 In *The Second Industrial Divide*, Sabel and his co-author, Michael Piore, are explicitly concerned with production trajectories rather than the technology–work nexus which Sabel covered in *Work and Politics*. However, their more synoptic perspective of general industrial change inevitably glosses over the sectoral specificities. These latter, I have argued, are central to understanding the factory as a socio-technical and economic institution. Despite the patent continuation of small-batch production in metalworking, Piore and Sabel present mass production as having achieved supremacy by the mid-twentieth century; albeit with important national variations which they ascribe to eclectic combinations of national traditions and conflicts, and to preservations of what the authors now counterpose as the 'craft' paradigm. See Piore and Sabel (1984, pp. 47, 133–64). For critical analyses of the conceptual polarity of mass production vs. flexible specialisation see Jones 1990, and Smith 1987.

2 In the early 1970s between 70 per cent and 85 per cent of US industrial output was estimated to be made by batch production methods. 35 per cent of all US manufacturing plant was for batch production (Kamrarny 1974, p. 282; General Accounting Office n.d., pp. i–ii) Littler refers to an unattributed estimate of up to 80 per cent of components being produced in batches of less than 1,000 (Littler 1983, p. 142). Kaplinsky (1984, p. 131) cites an estimate of 75 per cent of batch sizes being less than 100.

3 Williamson was presumably referring to aggregate time on several machines. Figures for the time individual machines cut metal may be between 30 per cent and 50 per cent of their availability (Tinker n.d., McKeown 1981).

4 Melman cites figures showing the US military budget composing less than 5 per cent of US GNP but 46 per cent of producers' fixed capital formation (Melman 1983, p. 261).

5 The internal organisational principles are almost identical to the organisation

of tasks and work flow in Japanese-inspired, 'just-in-time' production cells. See Turnbull 1988.

6 SYSTEM 24 has been the subject of a long-standing legal action brought by the Molins company against various US suppliers of FMS. It is alleged that the SYSTEM 24 patent covers the same system properties as the later FMS (cf. p. 135, ch. 6 below).

7 See Rosenberg's (1976, p. 73) criticism. For an outline of this perspective see Noble 1984, pp. 144–7.

8 Noble's account describes only one competitor to NC, the operator-programmed 'record-playback' system, and overlooks the rival techno-organisational systems of GT and integrated cells examined here. In fact, types of record-playback control were later incorporated into industrial robots. On this point and the technical limitations of record-playback see Sabel 1982, pp. 64–5. In fact, analogous operator-determined programming is now widely available on some models of CNC machine-tool.

9 For a fuller discussion of claimed and recorded advantages of NC operations cf. Jones 1983.

10 Although an expansive literature on the nature of skill has recently developed – partly in the wake of Braverman and the labour process debates – a systematic and socially grounded conceptualisation is still conspicuously absent. For some outline treatments cf. Adler and Borys 1986; Grootings, Jones and Scott 1989; Spenner 1983; Rolfe 1986; Libetta 1988.

11 A thorough discussion of the boundaries between technical and social, and the related problems of technological vs. social determinism raises fundamental issues and reference to an enormous literature. Mostly, however, these debates are concerned with the effect on work and social organisation. The issue in this present account is precisely the opposite causation: how the social factors shape, or more strictly interact with, the technology. For a recent interesting attempt to conceptualise the technology–work nexus, which parallels the above account, see McLoughlin and Clark 1988. For a variety of perspectives on technologies as subject to 'social shaping' see Mackenzie and Wacjman 1985.

5 *Numerical control, work organisation and societal institutions*

1 Sabel shows awareness of this omission in his treatment of the collective perspectives and strategies of interest groups. In this particular instance he recommends explicitly comparative analysis; particularly the matching case studies already undertaken by Sorge et al. (Sabel 1982, pp. 191–2). Elsewhere, however, the contradiction betwen universal production strategies and societally specific, though methodologically universal, social institutions is glossed over. The most articulate representatives of the latter approach are the French Aix group and their collaborators, such as Sorge et al. See Sabel 1982, p. 24, and the assessment of the significance of the Aix group programme in Rose 1985.

2 Most of the following discussion of British developments draws on previously published research findings in Jones 1982, 1985a and 1985b.

3 AEUW Circular to District Secretaries no. D.7/81.

4 This estimate assumes that the ratio of 26 per cent programmed by AEUW, 38 per cent by TASS or staff union members, 2 per cent by 'manufacturers,

suppliers and agents', and 6 per cent by members of other unions' and organisations was replicated in the 28 per cent of cases where responsibility was not identified.

5 Because the numbers of FMS are small, in both relative and absolute terms, these proportions can be taken as relating overwhelmingly to NC and CNC.

6 In a more unusual case, Giordano describes how unionised planning staff fought to retain overall responsibility when new CNC machinery allowed direct, though simplified, programming by operators (Giordano 1992).

7 The following exposition is a synthesis of the accounts in: Regalia et al. 1978; Romagnoli and Della Rocca 1982; Trentin 1981.

8 For an exemplary statement of this position cf. Il Manifesto in Gorz 1976.

9 B. Jones: Sasib interviews, June 1985. It is interesting to compare Murray's interpretation that the FIOM regional union leadership resisted the radical work reform scheme proposed by the *consiglia di fabbrica*, with the account of Garibaldo – FIOM provincial, then regional, secretary. The latter specifically singles out the mistrust of the technicians with egalitarian workgroup organisation as contributing to the ultimate lack of success at GD; Garibaldo 1988, pp. 136–7.

Overview to Part III

1 'Mechanical parts that are designed with a computer-based graphic design system should not require human intervention to create the program required to machine those parts . . . An automated knowledge based system will drive down NC programming, scrap cost, while productivity and quality will go up' (N. H. Woyak, Manager, Allen-Bradley Inc., cited in Dempsey and Pearce 1994, p. 1).

'What is sought is a reduction in the number of man-hours needed to achieve a changeover, perhaps combined with a de-skilling of the operations . . .' (McIntosh et al. 1994, p. 491).

6 *The cybernated factory and the American dream*

1 *American Machinist*, July 1983, p. 35. The patent award to Molins was subsequently revoked.

2 Author's interview with Nat Cooke, 10 August 1984.

3 Author's interview with Nat Cooke, 10 August 1984.

4 Miller and O'Leary (1994) describe computer-integrated automation as two of four pillars of the 'Plant With A Future' corporate manufacturing strategy of Caterpillar, one of whose FMSs is described in the present chapter. According to Miller and O'Leary, in Caterpillar's visionary ideology 'consolidation, simplification, automation and integration' were 'prescribed as the steps to be gone through in every facility and, as far as practicable, in that order' (Miller and O'Leary 1994, p. 31). What is unclear is whether the strategic vison is the source of specific technological changes, or whether they are rationalised, after the event, as outcomes of the corporate 'vision'. Miller and O'Leary's account (1994, p. 33) does not rule out technological developments such as FMS actually being formulated at plant level and then tailored to meet corporate-level

criteria; as Thomas (1994) suggests that local managers manipulate corporate investment criteria. From my visits to Caterpillar it was clear that local FMS developments were already in train *before* the formalisation of the corporate 'Plant With A Future' grand strategy.

5 For a fuller description of this phenomenon see Jones 1985c and, for a more complex case with NC and CNC, see Giordano 1992.

6 Further details of these cases are given in Jones 1989a and 1985c.

7 An American deviant: FMS at Alpha

1 This and the other names of Alpha staff are pseudonyms.

2 A similar tactic seems to have been adopted at the automobile plant FMS described in Thomas's study; although there the main aim is said to be allowing programming tasks into production workers' job classifications without setting precedents for the rest of the plant (Thomas 1994, pp. 197–8).

3 See Brody 1980; and for criticisms and an overview see Jones 1985a; Tolliday and Zeitin 1985.

8 Easy-peasy Japanesy: flexible automation in Japan

1 Conti and Warner have put forward a contrary interpretation of Taylorism in Japan. Because of the early appearance of Taylorist policies, in the textile and railway industries, and at Nippon Electric and Mitsubishi Heavy Electric Co., as well as the popularity of scientific management publications, Conti and Warner conclude that Taylorist influence began before the First World War and was established during the inter-war period. However, this interpretation is based upon the spread of discrete Taylorist *techniques* – principally time and motion studies. No case is made for the dominance of the Taylorist troika of redesigned jobs, time-studied payment-by-results and – most crucially – hierarchical planning and control in the metalworking industry. Indeed their qualification that Taylorism reached the auto industry only after the Second World War is consistent with the chronology suggested here (Conti and Warner 1993, pp. 4–7).

2 Contrasting British arrangements, in which – especially during the 1970s – shopfloor representatives are granted full-time trade union activity whilst being paid by the company, were associated with their 'incorporation' into managerial norms of conduct. See Hyman 1979 and Terry 1979.

3 Shopfloor skills in smaller machine-tool factories in the 1920s were at the same tacit levels as Britain and the USA a century earlier. The military complained that: 'They never used blueprints in making whatever machines they were making. They never thought of using gauges . . . They did all of their work by feel' (Toyasaki Minoru, *Nihon kikai kogyo no kiso kozo* (Tokyo: Nihon Hyoronsha 1941), cited in Yakamamura 1986).

4 This policy had unexpected, though short-term, successes in Okochi's Riken industrial group. Its main, and idiosyncratic, interpretation of scientific management was a Fordist strategy of specialising factories on mass production of single products using cheap, unskilled labour. See Cusumano 1989.

5 Although Nohara implies that, in general, skill levels are lower in the small firms (Nohara 1987, p. 22).
6 Most interviewing was in English, without interpreters wherever possible; to avoid their possible confusion of technical production and employment terms. Only with some of the employees of one FMS firm was this not possible. But in this case I was indebted to Jun Yamada who, as an industrial sociologist with a special interest in shopfloor society, and knowledge of Anglo-American debates, was able to question and interpret with great proficiency. All interviews were carried out in the summer of 1985. For comparative yardsticks I interviewed at two other firms: a small independent CNC-based machine shop with 150 machines planning an FMS, and a small industrial machinery supplier using conventional, plus two CNC machines. Another comparator is Masami Nomura's 1990, different, case study of a Japanese FMS (Nomura 1990).
7 Yamanote is the term for the original elite hill-side residential areas of Tokyo, but which literally means 'towards the mountains' ('Tokyo and London: Comparative Conceptions of the City' in Craig 1979, p. 91. For this was a new rural plant two hours from the congested industrial suburbs of Tokyo where the firm's first factory was located.).
8 In 1988 90 per cent of all students completed senior high school, compared to 77 per cent in the USA (Whitehill 1991). All the case study firms I visited, including the small press factory, required senior high school qualifications as a minimum.

9 Revolution from above: FMS in Britain

1 On the changing role of engineers in such firms, see Jones et al. 1993.
2 Most of the following details and figures derive from a re-analysis of Peter Scott's detailed fieldwork data: see Appendix. I am grateful to him for his advice on this evidence. Errors of interpretation are, of course, my responsibility.
3 For a contrary assessment of the severity of financial restrictions on FMS use and investment see Lee 1992, 1996.
4 For a fuller description of this typology see Jones 1988.
5 For a more detailed account of these events see Jones and Scott 1987.
6 Fifteen product families are made there, yet nine of these have 'little internal variation' (Scott 1986, p. 347). So a figure of seven part-types may be more representative of the versatility of this FMS; putting it amongst the group with lowest product variety.

10 The third Italy and technological dualism

1 For a review of these arguments see Perrow 1991.
2 British increases over the shorter period from 1976 to 1982 averaged 4,955 p.a. (*Metalworking Production* 1977). However, this British source – which records also metal-forming equipment, as well as metal-cutting machine-tools – estimates Italy's 82,566 NC/CNC total as 15.2 per cent of all machinery. This was double the ratio prevailing in Britain (*Metalworking Production* 1988,

p. 15).

3 But see Michelsons 1987 on the fuzziness of the boundary between FIAT and its sub-contractors; and Cressey and Jones 1991 on neo-Fordism within FIAT Auto.

4 44.38 per cent of all manufacturing employees in the region by the mid 1980s according to Rubini 1988.

5 For a fuller description of these services, albeit drawn mainly from the paragon case of the Modena province, see Best 1992, pp. 209–11.

6 In a broader 1991 survey of 286 Emilian artisan firms with 55 per cent in metalworking, only 18 per cent of all firms had any computerised production (Guagnini 1992, pp. 52, 60).

7 Interview with FLM Emilia Romagna officials, March 1985.

8 There was a trend to more functional specialisation between partners in some of Rieser and Franchi's sample.

9 Employment of technical workers on design and planning tasks has increased substantially, although they have only reached 10 per cent of all those in artisanal firms. But the numbers of production workers have also grown, while the *soci*, the artisans themselves, are fewer in number (personal communication from V. Rieser; see also Franchi and Rieser 1991), although variations also exist according to the different types of market relationships of the artisanal firm: (see note 8 above).

10 However, Vittorio Rieser now observes a few *conto proprio* proprietors differentiating their technical managerial tasks the better to develop and market their products. This may favour a more impersonal multi-unit legal grouping of enterprise establishments; which may remove the firm from the artisanal legal category.

11 Campione and Velocita are pseudonyms, used to avoid compromising informants who regarded some of the information provided as confidential.

12 This is Stephen Wood's interpretation of the flexible specialisation thesis (see Wood 1989, pp. 12–13) and the substance of an analogous advocacy for the small-firm network form of flexible specialisation (Best 1992).

11 Conclusion: the struggle continues

1 For an iconoclastic critique of the claims for the 'Lean Production' paradigm in the auto industry see Berggren 1993.

2 Cf. Kelleher's account of the British branch of one of the Japanese firms discussed in chapter 6 above (Kelleher 1993, ch. 5).

Bibliography

Abegglen, J. C. (1958) *The Japanese Factory*, Glencoe, Illinois: The Free Press.
Abernathy, W. J. (1978) *The Productivity Dilemma*, Baltimore: Johns Hopkins University Press.
ACARD (Advisory Council for Applied Research and Development) (1983) *New Opportunities in Manufacturing*, London: HMSO.
Accornero, A. (ed.) (1976) *Problemi del movimento sindacale in Italia 1943–1973, Annali, Anno Sedecismo 1974–75*, Milano: Fondazione Giangiacomo Feltrinelli.
Adler, P. S. (1990) 'Managing High-Tech Processes: The Challenge of CAD/CAM', in M. A. von Glinow and S. A. Mohrman (eds.) *Managing Complexity in High Technology Organizations*, New York: Oxford.
Adler, P. and Borys, B. (1986) 'Automation and Work: The Machine Tool Case', *mimeo* Stanford University: Department of Industrial Engineering and Engineering Management.
Aglietta, M. (1979) *A Theory of Capitalist Regulation*, London: New Left Books.
Althusser, L. and Balibar, E. (1970) *Reading Capital*, London: New Left Books.
American Machinist (1967) 'New Wave in Manufacturing', *Special Report* 607, 11 September.
(1983) 'Manufacturing Systems', July, pp. 35–7.
(1985) 'Japan's Builders Embrace FMSs', February, pp. 83–8.
Amin, A. (1988) 'The Flexible Small Firm in Italy: Myths and Realities', *mimeo* Newcastle University: Centre for Urban and Regional Development Studies.

(1991) 'Post-Fordism: Models, Fantasies and Phantoms of Transition', in A. Amin (ed.) *Post-Fordism: A Reader*, Oxford: Blackwell.

Anglo-American Council on Productivity (1951) 'Productivity Team Report: Valves, Steel, Iron and Non-ferrous', London: Anglo-American Council on Productivity.

(1952) 'Final Report of the Council', London: Anglo-American Council on Productivity.

(1953) 'Productivity Team Report: Metalworking Machine Tools', London: Anglo-American Council on Productivity.

Arnold, H. L. and Faurote, F. L. (1915) *Ford Methods and Ford Shops*, New York: The Engineering Magazine Company.

Ascari, R., di Domenica, F. and Lama., R. (1984) *L'innovazione tecnologica, L'evoluzione dell'organizzazione del lavoro e delle professionalità nelle imprese metalmeccaniche Bolognesi*, Bologna: CFP.

Ayusawa, I. F. (1966) *A History of Labour in Modern Japan*, Honolulu: East–West Centre Press.

Baglioni, M. (1986) 'Technological Modernisation in Italy: Government Initiatives in the Eighties', paper to the EGOS-AWG on Trade Union Research conference: 'Trade Unions, New Technology and Industrial Democracy', University of Warwick, 6–8 June.

Barisi, G. (1980) 'Le notion de "professionalità" pour les syndicats en Italie', *Colloque Politique d'Emploi et Rapports Sociaux du Travail*, Paris: Université de Paris VII.

Barkan, J. (1986) *Visions of Emancipation: Italian Workers Since 1945*, New York: Praeger.

Batstone, E., Gourlay, S., Levie, H. and Moore, R. (1987) *New Technology and the Process of Labour Regulation*, Oxford: Clarendon Press.

Baxter, A. (1991) 'Confusion over the Levels of Integration and Flexibility' Computers in Manufacturing Survey, *Financial Times*, 14 May.

Bell, R. M. (1972) *Changing Technology and Manpower Requirements in the Engineering Industry* Watford: Sussex University Press/EITB.

Bennati, R. (1990) 'L'analisi del lavoro. Una ipotesi di percorso', in O. Marchisio (ed.).

Berggren (1993) 'Lean Production – The End of History?' *Work, Employment and Society*, 7 (2), pp. 163–88.

Bessant, J. (1991) *Managing Advanced Manufacturing Technology: The Challenge of the Fifth Wave*, Oxford: Blackwell.

Bessant, J., Levy P., Ley, C. and Tranfield, D. (1992) 'Coping With Chaos: Designing the Organisation for Factory 2000', in Institution of Electrical Engineers, *Proceedings of the Third International Conference on Factory 2000 – Advanced Factory Automation*, London: Institution of Electrical Engineers, pp. 6–11.

Best, M. (1992) *The New Competition: Institutions of Industrial Restructuring*, Oxford: Polity.

Bianchi, G. and Lauzi, G. (eds.) (1981) *I metalmeccanici. Documenti per una storia del FIOM*, Bari: De Donato.

Blackburn, P., Coombs, R. and Green, K. (1985) *Technology, Economic Growth and the Labour Process*, London: Macmillan.

Blumberg, M. and Gerwin, D. (1981) 'Coping With Advanced Manufacturing Technology', IIM/LMP Discussion paper 81–12, Berlin: Wissenschaftszentrum.

Boston Consulting Group (1985) 'A Strategic Study of the Machine Tool Industry', Report for the Commission of the European Communities, Dusseldorf/Munich/Paris/London: Boston Consulting Group.

Brackenbury, H. I. (1907) *The Engineer*, 30 August, pp. 220–1

Braverman, H. (1975) *Labor and Monopoly Capital*, New York: Monthly Review Press.

Briefs, U., Ciborra, C. and Schneider, L. (eds.) (1983) *Systems Design: For, With and By the Users*, London/Amsterdam: North Holland Publishing.

Bright, J. (1958) *Automation and Management*, Boston, MA: Harvard University Press.

British Institute of Management (1956) *Outline of Work Study*, London: British Institute of Management.

British Management Data Foundation (1981) *The Management of Automation*, London: British Management Data Foundation.

British Productivity Council (1953) *Productivity Team Report: Metalworking Machine Tools*, London: BPC.

(1954) *Industrial Engineering*, London: British Productivity Council.

(1953) *Productivity Team Report: Metalworking Machine Tools*, London: BPC.

Brivio, O. (1985) *Impatto dello svilluppo tecnologico sull' organizzazione del lavoro nella produzione meccanica in Italia*, Milan: MAIN.

Brody, D. (1980) *Workers in Industrial America*, New York: Oxford University Press.

Broehl, W. G. (1959) *Precision Valley, the Machine Tool Companies of Springfield Vermont*, Engelwood Cliffs, New Jersey: Prentice-Hall.

Brown, R. K. (1992) *Understanding Industrial Organisations*, London: Routledge.

Brown, W. (1962) *Piecework Bargaining*, London: Heinemann Educational.

Brusco, S. (1982) 'The Emilian Model: Productive Decentralisation and Social Integration', *Cambridge Journal of Economics*, 6 (2).

Brusco, S. and Righi, E. (1985) 'Local Government, Industrial Policy and Social Consensus', OECD/Italy Seminar 'Opportunities for Urban Economic Development', Venice.

Brusco, S. and Sabel, B. (1981) 'Artisan Production and Economic Growth', in F. Wilkinson (ed.).

Buchanan, D. (1983) 'Technological Imperatives and Strategic Choice', in G. Winch (ed.).

Burawoy, M. (1979) *Manufacturing Consent: Changes in the Labour Process Under Monopoly Capitalism*, Chicago: Chicago University Press.

Burnes, B. (1984) 'Factors Affecting the Introduction and Use of Computer Numerically Controlled Machine Tools', in T. Lupton (ed.).

Burns, N. D., and Backhouse, C. J. (1994) 'Driving the Change Programme in a Manufacturing Business', in Institution of Electrical Engineers: *Proceedings of the Fourth International Conference on Factory 2000 – Advanced Factory*

Automation London: Institution of Electrical Engineers, pp. 479–83.

Burns, T. and Stalker, G. M. (1961) *The Management of Innovation*, London: Tavistock.

Business Week (1981) Special Report: 'The Speedup in Automation', 3 August, pp. 58–67.

Capecchi, V. (1984) 'Tendenze del artigianato di produzione nella provincia di Bologna', conference on: Problemi dello svilluppo industriale e dell' occupazione a Bologna, Commune di Bologna.

(1989) 'Changing Patterns of Small Entrepreneurship', in B. Strumpel (ed.).

Capecchi, V. et al. (1978) *La piccola impresa nell' economia Italiana*, Bari: De Donato.

Capes, P. (1985) 'Automated Manufacture without FMS', *Metalworking Production*, 129 (11), pp. 66–70.

Carew, A. (1987) *Labour Under the Marshall Plan*, Manchester: Manchester University Press.

Carter, P. (1976) *Payment by Results in W. D. & H. O. Wills, mimeo*: University of Bath, Library.

Cavestro, W. (1989) 'Automation, New Technology and Work Content', in S. Wood (ed.).

Cazzaniga Francesetti, D., Rebessi, P. L. and Reiser V. (1988) *Svillupo e limiti dei sistemi flessibili di produzione*, Milano: Franco Angeli.

Cecchini, P. et al. (1988) *The European Challenge: 1992 – The Benefits of a Single Market*, Aldershot: Wildwood House.

Cella, G. P. and Treu, T. (eds.) (1982) *Relazioni industriali. Manuali per l'analisi della esperienza italiana*, Bologna: Il Mulino.

CGIL (1986) *L'impresa artigiana: ruolo e prospettive. Una nuova politica sindicale e contrattuale nel settore*, Roma: Ediesse.

Child, J. and Mansfield, R. (1972) 'Technology, Size and Organisation Structure', *Sociology*, 6 (3).

CISIM (1952) *Problemi Economici ed industriali delle Industriale Meccaniche Italiana*, Stanford Research Institute: CISIM.

Clark, R. (1979) *The Japanese Company*, New Haven, CN: Yale University Press.

Clegg, S. and Dunkerley, D. (eds.) (1981) *International Yearbook of Organisational Studies*, London: Routledge.

CNR, UCIMU, Politecnico (1987) *Macchine per produrre macchine*, Milano: Edizione Sole 24 ore.

Cole, R. E. (1971) *Japanese Blue Collar*, Berkeley: University of California Press.

(1979) *Work, Mobility and Participation*, Berkeley: University of California Press.

Collins, H. M. (1990) *Artificial Experts*, Cambridge, MA: MIT Press.

Collins, O., Dalton, M. and Roy, D. (1946) 'Restriction of Output and Social Cleavage in Industry', *Applied Anthropology*, 14 (1).

Conference of Socialist Economists (1976) *Conference of Socialist Economists*, pamphlet no. 1, London: Stage One Books.

Conti, R. F. and Warner, M. (1993) 'Taylorism, New Technology and JIT Systems in Japan', *New Technology, Work and Employment*, 8 (1).

Cook, N. (1975) 'Computer-Managed Parts Manufacture', *Scientific American*, February, pp. 23–30.

Craig, A. M. (ed.) (1979) *Japan: A Comparative View*, Princeton: Princeton University Press.

Cressey, P. and Jones, B. (1991) 'Business Strategy and the Human Resource: The Last Link in the Chain? Banks and Cars in European Perspective', paper for the Open University Course in Human Resource Strategy, *mimeo*: Bath University: School of Social Sciences.

Cressey, P. and Jones, B. (eds.) (1996) *Work and Employment in European Society: A New Convergence?*, London: Routledge.

Crouch, C. and Pizzorno, A. (eds.) (1978) *The Resurgence of Class Conflict in Western Europe Since 1968*, vol. I, *National Studies*, London: Macmillan.

Currie, R. M. (1954) 'The Development and Scope of Work Study', *Proceedings of the Institution of Mechanical Engineers*, 168 (25).

Cusamano, M. A. (1989) 'Scientific Industry Strategy, Technology and Entrepreneurship in Prewar Japan', in W. D. Wray (ed.).

Daito, E. (1979) 'Management and Labor: The Evolution of Employer–Employee Relations in the Course of Industrial Development', in K. Nakagawa (ed.).

D'Attore, P. P. (1985) 'Anche noi possiamo essere prosperi. Auiti ERP e politiche della produttivita negli anni cinquanta', *Quaderni Storici*, 58/20 (1).

Davison, J. P., Florence, P. S., Gay, B. and Ross, N. S. (1958) *Productivity and Economic Incentives*, London: George Allen and Unwin.

Della Rocca, G. (1976) 'L'offensiva politica dell' imprenditorialità nelle fabbriche', in A. Accornero (ed.).

Dempsey, P. A. (1983) 'New Corporate Perspectives on FMS', in K. Rathmill (ed.) *Proceedings of the Second International Conference on Flexible Manufacturing Systems*, Kempston: IFS/North Holland.

Dempsey, P. A., and Pearce, D. W. (1994) 'CAM Tools for 2000 – A Manager's View', in *Proceedings of the Fourth International Conference on Factory 2000 – Advanced Factory Automation*, London: Institution of Electrical Engineers, pp. 1–4.

Department of Trade and Industry (1984) *Flexible Manufacturing Systems: Information for Applicants*, London: Department of Trade and Industry.

Devinat, P. (1927) *Scientific Management in Europe*, Geneva: ILO Studies and Reports, Series B, no. 17.

Dina, A. (1986) 'La fabbrica automatica e l'organizzazione del lavoro', in FIOM/CGIL.

Dodgson, M. (1985) *Advanced Manufacturing Technology in the Small Firm*, London: Technical Change Centre.

Dore, R. (1973) *British Factory – Japanese Factory*, Berkeley: University of California Press.

Dosi, G. (1982) 'Technological Paradigms and Technological Trajectories: A Suggested Interpretation of the Determinants of Technological Change', *Research Policy*, 11, pp. 147–62.

Drudi, I. (1986) 'Qualche considerazione sulle imprese artigiane metalmeccaniche emiliano-romagnole alle luce dei dati censuari 1971–1981', in CGIL.

ECIPAR/FNAM (1984) *Materiali per il progetto di sviluppo e di qualificazione del'artigianato*, Bologna: Confederazione Nazionale dell' Artigianato.

Edwards, P., Hall, M., Marginson, P., Sisson, K., Waddington, J. and Winchester, D. (1992) 'Still Muddling Through', in A. Ferner and R. Hyman (eds.).
Elbaum, B. and Lazonick, W. (eds.) (1986) *The Decline of the British Economy*, Oxford: Clarendon Press.
Factory Management and Maintenance (1957), New York: McGraw-Hill Publishing, October.
Fadem, J. (1976) 'Fitting Computer-Aided Technology to Workplace Requirements: An Example', *Proceedings of the NC Society*, March, pp. 9–16.
(1986) 'Beware the Hi-Tec Fix', *New Management*, 4 (2), pp. 40–4.
Fairbairn, W. (1861) *Treatise on Mills and Millwork. Part I. On The Principles of Mechanism and Prime Movers*, London: Spottiswood & Co.
Ferner, A. and Hyman, R. (eds.) (1992) *Industrial Relations in the New Europe*, Oxford: Blackwell.
Filippucci, C. (1984) *L'artigianto in Emilia Romagna. Un indagine statistica*, Bologna: CLUEB.
Filippucci, C. and Lugli, L. (1985) *I servizi per le industrie in un sistema di piccole e medie imprese*, Milan: Franco Angeli.
FIOM/CGIL (1985/6) *I lavoratori dentro le innovazioni tecnologiche: uomini, macchine, società*, Torino: Rosenberg & Sellier.
Ford, H. (1924) *My Life and Work*, New York: Doubleday.
Franchi, M. and Rieser, V. (1991) 'Le categorie sociologiche nell'analisi del distretto industriale: tra communità e razionalizzazione', *Stato e Mercato*, 33, December.
Fraser Smith, E. W. (1917) *NE Coast Institution of Shipbuilders and Engineers Transactions*, vol. 33.
Freeman, C. (1985) *Engineering and Vehicles (Technological Trends and Employment 4)*, Aldershot: Gower Press.
Friedman, D. (1988) *The Misunderstood Miracle*, Ithaca, NY: Cornell University Press.
Gallie, D. (1978) *In Search of the New Working Class*, Cambridge: Cambridge University Press.
Garibaldo, F. (1988) *Lavoro, Innovazione, Sindicato*, Genova: Costa & Nolan.
Gebhardt, A. and Hatzold, O. (1974) 'Numerically Controlled Machine Tools', in L. Nabseth and G. F. Ray (eds.).
General Accounting Office, United States Congress (n.d.) *Manufacturing Technology – A Changing Challenge to Improved Productivity*, Report to the Congress by the Comptroller General.
Gerwin, D. and Leung, T. K. (1980) 'The Organisational Impacts of Flexible Manufacturing Systems: Some Initial Findings', *Human Systems Management*, 1, pp. 237–46.
Giedion, S. (1948) *Mechanization Takes Command*, New York: W. W. Norton.
Gilbert, K. R. (1958) 'Machine Tools', in C. Singer et al. (eds.).
Giordano, L. (1992) *Beyond Taylorism*, London: Macmillan.
Giovannetti, E. and Zini, A. (1989) 'Appendix 5', in E. Goodman (ed.).
Gobbo, F. (ed.) (1989) *Distretti e sistemi produttivi alle soglie degli anni novanta*, Milan: Franco Angeli.

Goldhar, J. D. and Jellinek, M. (1983) 'Plan for Economies of Scope', *Harvard Business Review*, November–December.
Goodman, E. (ed.) (1989) *Small Firms and Industrial Districts in Italy*, London: Routledge.
Gordon, A. (1985) *The Evolution of Labor Relations in Japan Heavy Industry*, Cambridge, MA: Harvard East Asia Monographs.
Gorz, A. (ed.) (1976) *The Division of Labour*, Brighton: Harvester Press.
Gould, S. J. (1991) *Wonderful Life. The Burgess Shale and the Meaning of History*, London: Penguin.
Gould, W. B. (1984) *Japan's Reshaping of American Labor Law*, Cambridge, MA: MIT Press.
Graham, A. (1972) 'Industrial Policy', in W. Beckerman (ed.), *The Labour Government's Economic Record 1964–1970*, London: Duckworth.
Graham, M. B. W. and Rosenthal S. R. (1985) 'Flexible Manufacturing Systems Require Flexible People', Boston University: Manufacturing Roundtable Research Report Series.
Granick, D. (1962) *The European Executive*, London: Weidenfeld and Nicolson.
Grootings, P., Jones, B. and Scott, P. J. (1989) 'Mastering Metals – Problems in Analysing and Classifying "New" Technical Jobs in Metalworking', *Vocational Training*, 3, pp. 11–20.
Guagnini, M. (1992) *La diffusione di nuove tecnologie nelle imprese artigiane dell' Emilia Romagna*, Bologna: Assessorato Industria, Artigianato, Commercio e Cooperazione, Regione Emilia Romagna.
Guagnini, M. and Pignatti, O. (1991) *L'Artigianto in Emilia Romagna negli Anni '80*, Milano: Franco Angeli.
Hall, J. W. (1926) 'Joshua Field's Diary of a Tour in 1821 Through the Midlands', *Transactions of the Newcomen Society*, 6 (28).
Hall, P. A. (1986) 'The State and Economic Decline', in B. Elbaum and W. Lazonick (eds.).
Harrison, B. (1989) 'Concentration Without Centralisation: The Changing Morphology of the Small Firm Industrial Districts of the Third Italy', International Symposium: Problems of Local Employment under Structural Adjustment, Tokyo: National Institute of Employment and Vocational Research, 12–14 September.
Hartley, J. (ed.) (1984) *Flexible Automation in Japan*, Berlin: IFS Publications.
Haruo, S. (1992) 'Japanese Capitalism: The Irony of Success', *Japan Echo*, 19 (2).
Hazelhurst, R. J. (1967) 'Numerical Control and Changes in Job Skills', M.Sc dissertation, University of Birmingham.
Heine, H. (1986) 'Some Current Strategy Problems of Italian Trade Unions', in O. Jacobi et al. (eds.).
Herrigel, G. (1991) 'Industrial Order in the Machine Tool Industry – A Comparison of the United States and Germany', in E. Hildebrand (ed.).
Hildebrand, E. (ed.) (1991) *Betriebliche Sozialverfassung unter Veranderungsdruck*, Berlin: Wissenchaftszentrum fur Sozialforschung/Editions Sigma.
Hirschhorn, L. (1984) *Beyond Mechanisation: Work and Technology in a Post-Industrial Age*, Cambridge, MA: MIT Press.

Hirst, P. Q. (1989) 'The Politics of Industrial Policy' in P. Q. Hirst and J. Zeitlin (eds.).
Hirst, P. Q. and Zeitlin, J. (eds.) (1989) *Reversing Industrial Decline*, Leamington Spa: Berg Press.
Hogan, M. J. (1987) 'American Marshall Planners and the Search for a European Neo-Capitalism', *American Historical Review*, 90 (1), pp. 44–72.
Hounshell, D. (1984) *From the American System to Mass Production 1800–1932*, Baltimore: Johns Hopkins University Press.
Hunt, N. C. (1952) *Methods of Wage Payment in British Industry*, London: Isaac Pitman and Sons.
Hutchinson, G. (1979) 'Advanced Batch Machining Systems', paper to the Numerical Control Society Conference, March, Los Angeles, *mimeo*: Milwaukee: University of Wisconsin–Milwaukee.
Hutton, G. (1953) *We Too Can Prosper*, London: George Allen and Unwin.
Hyman, R. (1979) 'The Politics of Workplace Trade Unionism', *Capital and Class*, 8.
(1988) 'Flexible Specialisation: Miracle or Myth?', in R. Hyman and W. Streeck (eds.).
Hyman, R. and Streeck, W. (eds.)(1988) *New Technology and Industrial Relations*, Oxford: Basil Blackwell.
IIAE (Incorporated Institution of Automobile Engineers) (1907) 'Works Organisation', pp. 109–52.
Il Manifesto (1976) 'Challenging the Role of Technical Experts', in A. Gorz (ed.).
Inagami, T. (1983) *Labour Management Communication at the Workshop Level*, Tokyo: Japan Institute of Labour, Japanese Industrial Relations Series, 11.
Institution of Engineers and Shipbuilders (1917) *Transactions*, 18 December.
Iwauchi, R. (1969) 'Adaptation to Technological Change', *Developing Economies*, 126, pp. 428–50.
Jacobi, O., Jessop, B., Kastendiek, H. and Regini, M. (eds.) (1986) *Technological Change, Rationalisation and Industrial Relations*, London: Croom Helm.
Jacobini, A. (1949) *L'Industria Meccanica Italiana*, Rome: Centro di Studi e Piani Tecnico-Economici.
Jacobsson, S. (1986) *Electronics and Industrial Policy. The Case of Computer Controlled Lathes*, London: George Allen and Unwin.
Jacoby, S. M. (1985) *Employing Bureaucracy. Managers, Unions and the Transformation of Work in American Industry, 1900–1945*, New York: Columbia University Press.
Jaikumar, R. (1984) 'Flexible Manufacturing Systems: A Managerial Perspective', Working Paper 1-74-078. Cambridge, MA: Harvard Business School, Division of Research.
(1986) 'Post-Industrial Manufacturing', *Harvard Business Review*, 64 (6), pp. 69–76.
Jefferys, J. B. (1945) *The Story of the Engineers*, London: Lawrence and Wishart.
Jerome, H. (1934) *Mechanization in Industry*, New York: National Bureau of Economic Research.
Jones, B. (1979) 'The Definition and Control of Occupational Tasks in the Process of Technological Change', Final Report to the Social Science Research Council, Boston Spa: British Library.

(1982) 'Destruction or Redistribution of Engineering Skills: The Case of Numerical Control', in S. Wood (ed.).

(1983), 'Technical, Organisational and Political Constraints on Operator Programming of NC Machines', in U. Briefs et al. (eds.).

(1984) 'Divisions of Labour and Distribution of Tacit Knowledge', in T. Martin (ed.).

(1985a) 'Controlling Production Work: Legal Administration and State Regulation in the US and British Aerospace Industries', in S. Tolliday and J. Zeitlin (eds.).

(1985b) 'Flexible Technologies and Inflexible Jobs: Impossible Dreams and Missed Opportunities', Technical Paper MM85-728, *World Congress on Human Factors in Automation*, Dearborn, Michigan: Society of Manufacturing Engineers.

(1985c) 'Work Opportunities in Flexible Automation', *End of Grant Report* (00232144) to Economic and Social Research Council, Swindon: ESRC.

(1988) 'Work and Flexible Automation in Britain: A Review of Developments and Possibilities', *Work, Employment and Society*, 2 (4).

(1989a) 'Flexible Automation and Factory Politics: Britain in Comparative Perspective', in P. Hirst and J. Zeitlin (eds.)

(1989b) 'When Certainty Fails: Inside the Factory of the Future', in S. Wood (ed.).

(1990) 'New Production Technology and Work Roles: A Paradox of Flexibility vs Strategic Control?', in R. Loveridge and M. Pitt (eds.).

Jones, B. and Rose, M. (1986) 'Re-Dividing Labour: Factory Politics and Work Re-organisation in the Current Industrial Transition', in K. Purcell et al. (editors).

Jones, B. and Saren, M. (1990) 'Politics and Institutions in Small Business Development. Comparing Britain and Italy', *Labour and Society*, 15 (3), pp. 287–99.

Jones, B. and Scott, P. J. (1987) 'Flexible Manufacturing Systems in Britain and the USA', *New Technology, Work and Employment*, 2 (1).

Jones, B. and Wood, S. (1984) 'Qualifications Tacites, Division du Travail et Nouvelles Technologies', *Sociologie du Travail*, 26 (4).

Jones, B., Bolton, B., Bramley, A. and Scott, P. (1993) 'Human Resource Elite or Technical Labourers? Professional Engineers in Transnational Businesses', *Human Resource Management Journal*, 4 (1).

Kamramy, N. M. (1974) 'Technology: Measuring the Socio-Economic Impact of Manufacturing Automation', *Socio-Economic Planning*, 8, pp. 281–92.

Kaplinsky, R. (1984) *Automation*, Harlow: Longman.

Kearney and Trecker Corporation (n.d.) *Manufacturing Systems Application Workbook* (Managers Edition), Milwaukee: Kearney and Trecker.

Kelleher, M. (1993) 'Skill Shortages and Work Organisation in British Manufacturing', University of Bath: Ph.D dissertation.

Kelley, M. (1984) 'Tasks and Tools. An Inquiry into the Relationship Between Tasks, Skills and Technology, with Reference to the Machining Labour Process', Ph.D. dissertation, Massachussets Institute of Technology, Sloane School of Management.

(1989) 'Alternative Forms of Work Organisation Under Programmable Automation', in S. Wood (ed.).

Kelly, J. (1982) *Scientific Management, Job Redesign and Work Performance*, London: Academic Press.

Kern, H. and Pichierri, A. (1996) 'The New Production Concepts in Germany and Italy', in P. Cressey and B. Jones (eds.).

Kern, H. and Schumann, M. (1984) *Das Ende der Arbeitsteilung? Rationalisierung in der industriellen Produktion*, Munich: Verlag C. H. Beck.

King, R. (1985) *The Industrial Geography of Italy*, London: Croom Helm.

Kluzer, S. and Patarozzi, M. (1987) 'FMS in Fiat Trattori', *Lito Newsletter*, no. 8–9.

Knights, D., Willmott, H. and Collinson, D. (eds.) (1985) *Job Redesign: Critical Perspectives on the Labour Process*, Aldershot: Gower Press.

Knights, D. and Willmott, H. (eds.) (1988) *New Technology and the Labour Process*, London: Macmillan.

Kochan, T. A., Katz, H. and McKersie, R. (1986) *The Transformation of American Industrial Relations*, New York: Basic Books.

Koike, K. (1987) 'Human Resource Development and Labor Management Relationships', in K. Yakamamura and Y. Yasuba (eds.).

Kuhn, T. S. (1962) *The Structure of Scientific Revolutions*, Chicago: University of Chicago Press.

Kumazawa, M. and Yamada, J. (1989) 'Jobs and Skills under the Lifelong Nenko Employment System', in S. Wood (ed.).

Lacey Ford, R. (1986) *Ford: The Men and the Machine*, London: Heinemann.

Lanchester, F. W. (1907) 'Discussion', *Proceedings of the Incorporated Institution of Automobile Engineers*, pp. 131–3, 1906/7.

Landes, D. S. (1969) *The Unbound Prometheus. Technological Change and Industrial Development in Western Europe from 1750 to the Present*, Cambridge: Cambridge University Press.

Lane, C. (1988) 'Industrial Change in Europe: The Pursuit of Flexible Specialisation in Britain and West Germany', *Work, Employment and Society*, 2 (2), pp. 141–68.

Layton, E. (1974) 'The Diffusion of Scientific Management and Mass Production from the US in the Twentieth Century', *Proceedings of the 14th International Congress in the History of Science*, Tokyo 1974, pp. 377–86.

Lazerson, M. (n.d.) 'Organisational Growth of Small Firms: An Outcome of Markets and Hierarchies?', *mimeo*, Department of Law, Fiesole: European University Institute.

Lazonick, W. H. (1979) 'Industrial Relations and Technical Change: The Case of the Self-Acting Mule', *Cambridge Journal of Economics*, 3, (3), 231–62.

Leadbeater, C. (1989) 'Manufacturing Efficiency, Added Value Emanating From Acronyms', *Financial Times*, 13 December.

Lee, B. (1992) 'The Cost Conflicts of Flexible Automation', Ph.D dissertation, University of Bath.

(1996) 'The Justification and Monitoring of Advanced Manufacturing Technology: An Empirical Study of 21 Installations of Flexible Manufacturing Systems', *Management Accounting Research*, 7, pp. 95–118.

Leonard, R. and Rathmill, K. (1977) 'Group Technology – A Restricted

Manufacturing Technology', *Chartered Mechanical Engineer*, 24 (9).
Levine, S. (1965) 'Labor Markets and Collective Bargaining' in W. L. Lockwood (ed.).
Levine, S. and Kawada, H.(1980) *Human Resources in Japanese Industrial Development*, Princeton: Princeton University Press.
Lewchuck, W. (1986) 'The Motor Vehicle Industry', in B. Elbaum and W. Lazonick (eds.).
Libetta, L. (1988) 'Tacit Knowledge and the Computerisation of Skills', Ph.D dissertation, University of Bath.
Lichtenstein, N. and Meyer, S. (eds.) (1985) *The American Automobile Industry: A Social History*, Champaign-Urbana IL: University of Illinois Press.
Littler, C. R. (1982) *The Development of the Labour Process in Capitalist Societies*, London: Heinemann Educational.
(1983) 'A History of 'New' Technology', in G. Winch (ed.).
Lockwood, W. L. (ed.) (1965) *The State and Economic Enterprise in Japan*, Princeton, NJ: Princeton University Press.
Loveridge, R. (1981) 'Business Strategy and Community Culture', in S. Clegg and D. Dunkerley (eds.).
Loveridge, R. and Pitt, M. (eds.) (1990) *The Strategic Management of Technological Innovation*, Chichester: John Wiley.
Lupton, T. (1967) Select Written Evidence 11, to UK Government, Royal Commission on Trade Unions and Employers' Associations, London: HMSO.
Lupton, T. (ed.) (1984) *Proceedings of the First International Conference on Human Factors in Manufacturing*, Bedford: IFS Publications.
Mackenzie, D. and Wacjman, J. (eds.) (1985) *The Social Shaping of Technology*, Milton Keynes: Open University Press.
Mahadeva, K. (1986) *Advanced Manufacturing Technology in the West Midlands. A Pilot Survey Final Report to SERC*, City of Birmingham Polytechnic: Department of Mechanical and Production Engineering.
Maier, C. S. (1970) 'Between Taylorism and Technocracy', *Contemporary History*, 5 (2), pp. 27–62.
Mansfield, E. (1971) 'The Diffusion of a Major Manufacturing Innovation', in J. Rapoport et al. (eds.)
Marchisio, O. (ed.) (1990) *Frammenti di sindicato*, Milano: Franco Angeli.
Marglin, S. A. (1976) 'What Do Bosses Do?', in A. Gorz (ed.).
Marsh, R. M. and Mannari, H. (1988) *Organizational Change in Japanese Factories*, Greenwich, CN: JAI Press.
Martin, P. (1907) 'Works Organisation', *Proceedings of the Incorporated Institution of Automobile Engineers*, pp. 109–28, 1906/7.
Martin, T. (ed.)(1984) *Design of Work in Automated Production Systems*, Oxford: Pergamon Press.
Marx, K. (1967) *Capital. A Critique of Political Economy*, vol. I, New York: International Publishers.
Marx, K. and Engels, F. (1986) *Collected Works*,vol. XXVIII, London: Lawrence and Wishart.
Matera, P. (1985) 'Small Is Not Beautiful. Decentralised Production and the

Underground Economy in Italy', *Radical America*, pp. 67–76.
Mathias, P. and Postan, M. M. (1978) *The Cambridge Economic History of Europe*, vol. III, *The Industrial Economies Capital Labour and Enterprise*, Part 2, *The United States, Japan and Russia*, Cambridge: Cambridge University Press.
Maurice, M., Sellier, F. and Silvestre, J. J. (1986) *The Social Foundations of Economic Power*, Cambridge, MA: MIT Press.
Mayfield J. (1979) 'Factory of the Future Researched', *Aviation Week and Space Technology*, 5 March.
McIntosh, R. I., Culley, S. J., Gest, G. B. and Mileham, A. R. (1994) 'Achieving and Optimising Flexible Manufacturing Performance on Existing Manufacturing Systems', in Institution of Electrical Engineers, *Proceedings of the Fourth International Conference on Factory 2000 – Advanced Factory Automation*, London: Institution of Electrical Engineers, pp. 455–491.
McKeown, P. (1981) 'Automation in Engineering Manufacturing – The World Scene', in British Management Data Foundation.
McLoughlin, I. and Clark, J. (1988) *Technological Change at Work*, Milton Keynes: Open University Press.
Melman, S. (1983) *Profits Without Production*, New York: Alfred Knopf.
Melossi, D. (1977) *Lotta di classe nel piano di lavoro*, Roma: Editrice Sindicale Italiana.
Merkle, J. A. (1980) *Management and Ideology: The Legacy of the International Scientific Management Movement*, Berkeley/Los Angeles: University of California Press.
Meyer, S. (1981) *The Five Dollar Day: Labour, Management, and Social Control in The Ford Motor Company*, Albany: State University of New York.
Michelsons, A. (1987) 'Turin Between Fordism and Flexible Specialisation. Industrial Structure and Social Change 1970–1985', D.Phil, University of Cambridge.
Miller, P. and O'Leary, T. (1994) 'Accounting, "Economic Citizenship" and the Spatial Reordering of Manufacture', *Accounting, Organisations and Society*, 19 (1), pp. 15–43.
Millward, N., Stevens, M., Smart, D. and Hawkes, W. R. (1992) *Workplace Industrial Relations in Transition: The ED/ESRC/PSI/ACAS Surveys*, Aldershot: Dartmouth.
Momigliani, F. (1986) 'Programmi nazionali di ricerca approvati in base alla lege 46/1982', in CER-IRS, *La legislazione sulla politica industriale: obbietivi, strumenti, risultati*, vol. III, Bologna: Il Mulino.
Murray, F. (1984) 'Industrial Restructuring and Working Class Politics: A Study of the Decentralisation of Production in the Bologna Engineering Industry', Ph.D dissertation, University of Bristol.
(1988) 'The Decentralisation of Production – the Decline of the Mass-Collective Worker', in R. E. Pahl (ed.).
Murray, R. (1985) 'Benetton Britain: The New Economic Order', *Marxism Today*, 29 November.
Musson, A. E. (1960) 'An Early Engineering Firm: Peel, Williams & Co of Manchester', *Business History*, 3, pp. 8–18.
Nabseth, L. and Ray, G. F. (eds.) (1974) *The Diffusion of New Industrial Processes:*

An International Study, Cambridge: Cambridge University Press.
Nakagawa, K (ed.) (1979) *Labor and Management*, Tokyo: University of Tokyo Press.
Nakase, T. (1979) 'The Introduction of Scientific Management in Japan and its Characteristics – Case Studies in the Sumitomo Zaibatsu', in K. Nakagawa (ed.).
National Engineering Laboratory (1978) *Automated Small Batch Production*, Glasgow: National Engineering Laboratory.
Nelson, D. (1980) *Frederick W. Taylor and the Rise of Scientific Management*, London/Madison: University of Wisconsin Press.
Nelson, R. R. (1987) *Understanding Technical Change as an Evolutionary Process*, Amsterdam/Oxford: New Holland.
Newman, S. T. and Bell, R. (1992) 'The Modelling of Multi-cell Flexible Machining Facilities', in Institution of Electrical Engineers, *Proceedings of the Third International Conference on Factory 2000 – Advanced Factory Automation*, London: Institution of Electrical Engineers, pp. 285–90.
Nichols, T. (1986) *The British Worker Question*, London: Routledge.
Noble, D. (1984) *Forces of Production: A Social History of Industrial Automation*, New York: Alfred Knopf.
Nohara, H. (1989) 'Technological Innovation, Industrial Dynamic and the Transformation of Work: The Case of the Japanese Machine Tool Industry', paper presented to the conference on 'High Technology and Society in Japan and the Federal Republic of Germany', Wissenschaftszentrum, Berlin.
Nomura, M. (1990) 'Social Conditions for CIM in Japan. A Case Study on a Machine Tool Company', paper to the international conference on 'Company Social Constitutions under Pressure to Change', Berlin: Wissenschafszentrum.
Olling, G. (1978) 'Experts Look Ahead to the Day of the Full-Blown Computer-Integrated Factory', *IEEE Spectrum*, October, pp. 60–6.
Pahl, R. E. (ed.) (1988) *On Work*, Oxford: Blackwell.
Palloix, C. (1976) 'The Labour Process: from Fordism to neo-Fordism', in Conference of Socialist Economists.
Pasini, E. (1985) 'Primi risultati della "Indagine sulla introduzione di tecnologie informatiche nelle aziende meccaniche dell" Emilia Romagna', draft, Bologna: IRES.
Patrick, H. T. and Rohlen T. P. (1987) 'Small Scale Family Enterprises' in K. Yakamamura and Y. Yasuba (eds.).
Pelzel, J. C. (1979) 'Factory Life in Japan and China Today', in A. M. Craig (ed.).
Penn, R. and Wigzell, B. (n.d.) 'The Attitudes and Responses of Trade Unions to Technical Change: A Case Study of Maintenance Workers in the North-West of England', *mimeo*: University of Lancaster, Department of Sociology.
Perez, C. (1985) 'Micro-Electronics, Long Waves and World Structural Change: New Perspectives for Developing Countries', *World Development*, 13 (3), 441–63.
Perrow, C. (1991) 'A Society of Organisations', *Theory and Society*, 20, pp. 725–62.
Petre, P. (1985) 'How GE Bobbled the Factory of the Future', *Fortune*, 11 November, pp. 46–54.

Piore, M. J. and Sabel, C. F. (1984) *The Second Industrial Divide*, New York: Basic Books.

Pollard, H. (1965) *The Genesis of Modern Management*, London: Edward Arnold.

Pollert, A. (ed.) (1991) *Farewell to Flexibility*, Oxford: Blackwell.

Pullin, J. (1987) 'The "Engineer" Survey: Flexible Friends', *The Engineer*, 23–30 April.

Purcell, K., Wood, S., Waton, A. and Allen, S. (eds.) (1986) *The Changing Experience of Employment*, London: Macmillan.

Rader, L. T. (1969) 'Computers and People . . . Partners in the Total Manufacturing Systems of Tomorrow', *Machine and Tool Blue Book*, July, pp. 98–101.

Rampini, F. (1982) 'Interrogativi sui Quadri', *Rinascita*, 2 April.

Rapoport, J., Schnee, S., Wagner, S. and Hamburger, M. (eds.) (1971) *Research and Innovation in the Modern Corporation*, London: Macmillan.

Regalia, I., Regini, M. and Reyneri, E. (1978) 'Labour Conflicts and Industrial Relations in Italy', in C. Crouch and A. Pizzorno (eds.)

Rieser, V. (1986) 'Aspetti e immagini della professionalita nelle imprese artigiane e industriali. Un prima rapporto di ricerca', in CGIL, *Quaderni Storici*, 20 (1), April.

(1988) 'Impiegati e sindicato. Spunti dalle recenti ricerche tra i metalmeccanici', *Ex Machina*, 9, November–December.

Rieser, V. and Franchi, M. (1986) 'Innovazione tecnologica e mutamento organizzativo nel'impresa artigiana', Modena: SNAMM.

Roethlisberger, F. J. and Dickson W. J.(1964) *Management and the Worker*, New York: Wiley.

Roitman, D. B., Liker, J. K. and Roskies, E. (1986) 'Birthing a Factory of the Future: When is "All At Once" Too Much?', ITI paper 86–21, November, Ann Arbor, MI: Industrial Technology Institute.

Rolfe, H. (1986) 'Skill, De-Skilling and New Technology in the Non-Manual Labour Process', *New Technology Work and Employment*, 1 (1).

Roll, E. (1930) *An Early Experiment in Industrial Organisation. A History of the Firm of Boulton & Watt 1775–1805*, London: Longman.

Rollier, M. (1976) 'The Organisation of Work and Industrial Relations in the Italian Engineering Industry', *Labour and Society*, pp. 81–4.

(1986) 'Changes of Industrial Relations at FIAT', in O. Jacobi et al (eds.).

Rolt, L. T. C. (1965) *Tools for the Job. A Short History of Machine Tools*, London: Batsford.

Romagnoli, G. and Della Rocca, G. (1982) 'Il sindicato', in G. P. Cella and T. Treu.

Rose, M. J. (1985) 'Universalism, Culturalism and the Aix Group: Promise and Problems of a Societal Approach to Economic Institutions', *European Sociological Review*, 1 (1).

(1988) *Industrial Behaviour. Research and Control*, London: Penguin.

(1994) 'Trade Unions Retreat or Rally?' *Work, Employment & Society*, 7 (2), June.

Rose, M. J. and Jones, B. (1985) 'Managerial Strategy and Trade Union Response in Work Re-Organisation Schemes at Establishment Level', in D. Knights et al.

Rosenberg, N. (1976) *Perspectives on Technology*, Cambridge: Cambridge University Press.

Rowe, J. W. F. (1928) *Wages in Practice and Theory*, London: Routledge.
Roy, D. (1952) 'Quota Restriction and Goldbricking in a Machine Shop', *American Journal of Sociology*, 57, pp. 427–42.
Rubini, I. (1988) *A Strategy for Developing Small Manufacturing Business*, Bologna: Confederazione Nazionale dell' Artigianato.
Sabel, C. (1982) *Work and Politics*, Cambridge: Cambridge University Press.
Sabel, C. and Zeitlin, J. (1985) 'Historical Alternatives to Mass Production', *Past and Present*, 108.
Schonberger, R. J. (1982) *Japanese Manufacturing Techniques*, New York: The Free Press.
Schotters, F. A. (1947) 'Incentive Plans *vs* Measured Day Work', *A.M.A. Production Series*, no. 19, New York: American Management Association.
Scott, P. B. (1984) *The Robotics Revolution*, Oxford: Basil Blackwell.
Scott, P. J. (1986) 'Craft Skills in Flexible Manufacturing Systems', Ph.D dissertation, University of Bath.
Shaiken, H. (1984) *Work Transformed: Automation and Labour in The Computer Age*, New York: Holt, Rheinhart and Winston.
Shirai, T. (ed.) (1983) *Contemporary Industrial Relations in Japan*, Madison: University of Wisconsin Press.
Sim, R. M. (1984) 'Encouraging the Growth of Advanced Manufacturing Technology Through the DTI Support Schemes', in Institute of Electrical Engineers, *International Conference on Computer Aided Engineering Conference Publication no. 243*, London: IEE, pp. 135–8.
Singer, C. et al. (1958) *A History of Technology*, Oxford: Oxford University Press.
Small, B. (1992) 'Factory 2000 – Translating the Vision into Reality, Proceedings of Third International Conference on *Factory 2000 – Advanced Factory Automation*, London: Institution of Electrical Engineers
Smiles, S. (1863) *Industrial Biography. Iron Workers and Tool Makers*, 1967 reprint, Newton Abbot: David and Charles.
Smith, C. (1987) *Technical Workers. Class, Labour and Trade Unionism*, London: Macmillan.
(1991) 'From 1960s Automation to Flexible Specialisation a *déjà vu* of Technological Panaceas', in A. Pollert (ed.).
Smith, C., Child, J. and Rowlinson, M. (1991) *Re-Shaping Work: The Cadbury Experience*, Cambridge: Cambridge University Press.
Smuts, R. W. (1953) *European Impressions of the American Worker*, London: Geoffrey Cumberledge, Oxford University Press.
Sorge, A., Hartmann, G., Nicholas, I. and Warner, M. (1983) *Micro-electronics and Manpower in Manufacturing. Applications of Computer Numerical Control in Great Britain and West Germany*, Aldershot: Gower Press.
Spenner, K. (1983) 'Deciphering Prometheus: Temporal Change in the Skill Level of Work', *American Sociological Review*, 48 (6), pp. 834–7.
Stanovsky, C. (1981) 'Automation and Internal Labour Market Structure: A Study of the Caterpillar Tractor Company', M.Sc dissertation, Cambridge, MA: MIT Press.
Staples, G. (1987) 'Technology, Control, and the Social Organization of Work at

a British Hardware Firm, 1791–1891', *American Journal of Sociology*, 93 (1), pp 62–88.
Strumpel, B. (ed.) (1989) *Industrial Societies after the Stagnation of the 1970s – Taking Stock from an Inter-disciplinary Perspective*, Berlin/New York: De Gruyter.
Sward, K. (1948) *The Legend of Henry Ford*, New York: Russell and Russell.
Swords-Isherwood, N. and Senker, P. (1978) *Technological and Organisational Change in Machine Shops*, Report to the Engineering Industry Training Board, Brighton: Science Policy Research Unit, University of Sussex.
Taira, K. (1978) 'Factory Labour and the Industrial Revolution in Japan', in P. Mathias and M. M. Postan (eds.).
Taylor, F. W. (1947) *Scientific Management*, Westport, CN: Greenwood Press, 1972 reprint.
Terry, M. (1979) 'Shop Stewards: The Emergence of a Lay Elite?', University of Warwick: IRRU discussion paper.
Terry, M. and Edwards, P. K. (eds.) (1988) *Shopfloor Politics and Job Controls*, Oxford: Basil Blackwell.
Tewson, V. (1953) 'Relationship of the Trade Union Movement to Government and Industry', *Trade Unions Today*, London: British Institute of Management.
The Engineer (1907) 'Modern Machinery and its Future Development', 30 August 1907, pp. 220–1.
Thoburn, J. T. and Takashima, M. (1992) *Industrial Subcontracting in the UK and Japan*, Aldershot: Avebury.
Thomas, R. J. (1994) *What Machines Can't Do. Politics and Technology in the Industrial Enterprise*, Berkeley: University of California Press.
Thompson, H. and Scalpone R. (n.d.) 'Managing the Human Resource in the Factory of the Future', *mimeo*: Milwaukee: A. T. Kearney Inc.
Tinker, C. C. (n.d.) 'The Utilisation of Machine Tools', Research Paper no. 23, Machine Tool Industry Research Association.
Toffler, A. (1981) *The Third Wave*, New York: Bantam Books.
Togliatti, P. (1974) *Politica Nazionale e Emilia Rossa*, Rome: Editori Riuniti.
Tolliday, S. and Zeitin, J. (1985) 'Shop Floor Bargaining, Contract Unionism, and Job Control: An Anglo-American Comparison', in N. Lichtenstein and S. Meyer (eds.).
(1987) *The Automobile Industry and Its Workers: Between Fordism and Flexibility*, Cambridge: Cambridge University Press.
Tolliday, S. and Zeitlin, J. (eds.) (1985) *Shopfloor Bargaining and the State*, Cambridge: Cambridge University Press.
Trecker, J. L. (1947) 'The Economic Justification for Investment in Plant Assets', *A.M.A. Production Series*, no. 19, New York: American Management Association.
Trentin, B. (1981) 'La nuova cultura del movimento operaio', in G. Bianchi and G Lauzi (eds.).
Trigilia, C. (1986) 'Small-firm Development and Political Subcultures in Italy', *European Sociological Review*, 2 (3), December, pp. 161–75.
Trist E. L. *et al.* (1963) *Organisational Choice*, London: Tavistock Institute of Human Relations.

Tsuda, M. (1979) 'The Formation and Characteristics of the Work Group in Japan', in K. Nakagawa (ed.)
Turnbull, P. J. (1988) 'The Limits to Japanisation': Just-in-Time, Labour Relations and the UK Automotive Industry, *New Technology, Work & Employment*, 3 (1), pp. 7–20.
Turone, S. (1984) *Storia del Sindicato in Italia*, Bari: Laterza.
Twaalfhoven, E. and Hattori, T. (1982) *The Supporting Role of Small Japanese Enterprises*, Netherlands: Indives Research Institute.
UK Government (1968) *Royal Commission on Trade Unions and Employers Associations*, London: HMSO.
(1983) House of Commons Industry and Trade Committee, *Third Report, 1982–83 Machine Tools and Robotics*, London: HMSO.
Ure, A. (1967) *The Philosophy of Manufactures*, New York: Augustus M. Kelley.
Utili, G. (1989) 'Mutamenti organizzativi nei distretti industriali:osservazioni su due casi' in F. Gobbo (ed.).
van Helvoort, E. (1979) *The Japanese Working Man*, Tenterden, Kent: Paul Norbury.
Wainwright, C. E. R., Harrison, D. K. and Leonard R. (1992) 'Manufacturing Strategy – A Lesson in Application', in Institution of Electrical Engineers, *Proceedings of the Third International Conference on Factory 2000 – Advanced Factory Automation*, London: Institution of Electrical Engineers, pp. 118–23.
Watson, W. F. (1935) *Machines and Men: Memoirs of an Itinerant Mechanic*, London: George Allen and Unwin.
Wells, H. G. (1935) *The Time Machine*, 1993 Everyman edition, London: J. M. Dent.
Whitehill, A. M. (1991) *Japanese Management: Tradition and Transition*, London: Routledge.
Whittaker, D. H. (1990) *Managing Innovation: A study of British and American Factories*, Cambridge: Cambridge University Press.
Wiener, N. (1948) *Cybernetics*, Cambridge, MA: MIT Press.
Wild, R. (1975) *Work Organisation*, London: John Wiley.
Wilkinson, B. (1982) *The Shopfloor Politics of New Technology*, London: Heinemann.
Wilkinson, F. (ed.)(1981) *The Dynamics of Labour Market Segmentation*, Cambridge: Cambridge University Press.
Williams, J. R. (1983) 'Technological Evolution and Competitive Response', *Strategic Management Journal*, 4 (2), pp. 55–65.
Williams, K., Cutler, T., Williams, J. and Haslam, C. (1987) 'The End of Mass Production?' *Economy and Society*, 16 (3), pp. 405–39.
Williams, K., Haslam, C. and Williams, J. (1992) 'Ford *versus* 'Fordism'. The Beginning of Mass Production?', *Work, Employment and Society*, 6 (4), pp. 517–55.
Williamson, D. T. N. (1955) 'Computer-Controlled Machine Tools', in *The Automatic Factory What Does It Mean?*, Report of the conference at Margate, June 1955, London: Institution of Production Engineers, pp. 144–53.
(1967) 'New Wave in Manufacturing', *American Machinist Special Report*, 607, 11 September, pp. 143–54.
(1968) 'The Pattern of Batch Manufacture and its Influence on Machine Tool

Design', *Proceedings of the Institution of Mechanical Engineers*, Part I, vol. 182, pp. 870–93.
Wilson, F. (1988) 'Computer Numerical Control and Constraints', in D. Knights and H. Willmott.
Winch, G. (ed.) (1983) *Information Technology in Manufacturing Processes. Case Studies in Technological Change*, London: Rossendale.
Winner, L. (1977) *Autonomous Technology: Technics Out of Control as a Theme in Political Thought*, Cambridge MA: MIT Press.
Womack, J. P., Jones, D. T., and Roos, D. (1990) *The Machine that Changed the World*, New York: Macmillan.
Wood, S. (ed.) (1982) *The Degradation of Work*, London: Hutchinson.
Wood, S. (1988) 'Between Fordism and Flexibility? The Case of The US Car Industry', in R. Hyman and W. Streeck (eds.).
(1989a) 'The Japanese Management Model', *Work and Occupations*, 16 (4), pp. 446–60.
Wood, S. (ed.) (1989b) *The Transformation of Work?*, London: Unwin Hyman.
Woodward, J. (1980) *Industrial Organisation: Theory and Practice*, 2nd edn, Oxford: Oxford University Press.
Wray, W. D. (ed.) (1989) *Managing Industrial Enterprise. Cases from Japan's Prewar Experience*, Cambridge, MA: Harvard University Press.
Yakamamura, K. and Yasuba, Y. (eds.) (1987) *The Political Economy of Japan*, vol. I, *The Domestic Transformation*, Stanford, CA: Stanford University Press.
Yates, M. L. (1937) *Wages and Labour Conditions in British Industry*, London: MacDonald and Evans.
Young, S. with Lowe, A. V. (1974) *Intervention in the Mixed Economy*, London: Croom Helm.
Yulke, S. G. (1947) 'Incentives and Work Standard Grievances', A.M.A. Production Series, no. 19, New York: American Management Association.
Zeitlin, J. (1979) 'Craft Control and the Division of Labour: Engineers and Compositors in Britain 1890–1930', *Cambridge Journal of Economics*, 3 (3), pp. 263–74.
(1987) 'Between Flexibility and Mass Production: Product, Production and Labour Strategies in British Engineering 1880–1939', *mimeo*: Birkbeck College, University of London.
(1989) 'Local Industrial Strategies: Introduction', *Economy and Society*, 18 (4), pp. 367–373.
Zuboff, S. (1988) *In the Age of the Smart Machine*, Oxford: Heinemann Publishing.
Zygmont, J. (1986) 'Flexible Manufacturing Systems – Curing the Cure-All', *High Technology*, October, pp. 22–5.

Index